# Automation, Production Systems and Computer-Integrated Manufacturing

# Automation, Production Systems and Computer-Integrated Manufacturing

**Raymond Foster**

Larsen & Keller
www.larsen-keller.com

Automation, Production Systems and Computer-Integrated Manufacturing
Raymond Foster
ISBN: 978-1-64172-094-6 (Hardback)

**⊟ Larsen & Keller**

Published by Larsen and Keller Education,
5 Penn Plaza,
19th Floor,
New York, NY 10001, USA

**Cataloging-in-Publication Data**

Automation, production systems and computer-
integrated manufacturing / Raymond Foster.
    p. cm.
Includes bibliographical references and index.
ISBN 978-1-64172-094-6
1. Automation. 2. Production management. 3. Computer integrated manufacturing systems.
4. Industrial engineering. I. Foster, Raymond.
T59.5 .A98 2019
670.427--dc23

For more information regarding Larsen and Keller Education and its products, please visit the publisher's website www.larsen-keller.com

# Table of Contents

# Preface

Automation is the technology that is designed to function without human assistance. Various control systems are used for the operation of equipment used in factories, boilers, ships, aircraft, etc. Automation is achieved by integrating hydraulic, electrical, mechanical, pneumatic and electronic devices and computers. It results in labor, electricity cost and material cost saving. It also ensures improvement of quality, precision and accuracy. Computer–integrated manufacturing is the approach to the use of computers for controlling the production process. It allows the exchange of information between processes. It is used in multiple domains, such as in mechanical engineering, electronic design automation, industrial and production engineering, etc. This book unfolds the innovative aspects of automation, production systems and computer-integrated manufacturing which will be crucial for the holistic understanding of modern manufacturing. Most of the topics introduced herein cover new techniques and the applications of these processes. As this field is emerging at a rapid pace, the contents of this book will help the readers understand the modern concepts and applications of the subjects.

A detailed account of the significant topics covered in this book is provided below:

**Chapter 1**, Computer integrated manufacturing is an approach to manufacturing that is guided by the use of computers during the production process. This is an introductory chapter, which will discuss briefly all the significant aspects of computer integrated manufacturing and includes vital topics such as computer aided technologies and manufacturing, control theory, etc. **Chapter 2**, Automation involves the use of control systems for the efficient execution of processes and operation of equipment and machinery with reduced or minimal human intervention. The aim of this chapter is to explore the need for automation in the industrial sector and its status in the modern scenario. It includes vital topics such as industrial automation, control loop, programmable logic controller and human machine interface. **Chapter 3**, Automation refers to the technology by which processes are performed without human assistance. This chapter has been carefully written to provide an easy understanding of the varied aspects of automation in product design, such as computer aided design, boundary representation, computer aided industrial design architectural rendering, assembly modeling, 3D modeling, etc. **Chapter 4**, Production support machines are vital for industrial manufacturing processes for effective material handling, optimum production or for the execution of complex functions. Robots are vital production support machines. The topics elaborated in this chapter like industrial robots, articulated robots, SCARA robots, delta robots and Cartesian coordinate robots will help in providing a better perspective about such machines. **Chapter 5**, The rapid advancement in electronic and automation technologies has brought tremendous innovation in computer aided production. This emerging branch of engineering studies such innovations in design and manufacturing. All the diverse topics central to the understanding of computer aided production such as computer numerical control machine, cutter location direct numerical control, pencil milling, etc. have been extensively discussed in this chapter. **Chapter 6**, Enterprise resource planning (ERP) refers to the integrated management of business processes, achieved by a combination of software and technology. All the diverse aspects of enterprise resource planning have been carefully analyzed in this chapter, such as enterprise planning systems, ERP modeling, manufacturing resource planning, master data management, etc.

I would like to make a special mention of my publisher who considered me worthy of this opportunity and also supported me throughout the process. I would also like to thank the editing team at the back-end who extended their help whenever required.

**Raymond Foster**

# Computer Integrated Manufacturing

Computer integrated manufacturing is an approach to manufacturing that is guided by the use of computers during the production process. This is an introductory chapter, which will discuss briefly all the significant aspects of computer integrated manufacturing and includes vital topics such as computer aided technologies and manufacturing, control theory, etc.

## Manufacturing

Manufacturing is the processing of raw materials into finished goods through the use of tools and processes. Manufacturing is a value-adding process allowing businesses to sell finished products at a premium over the value of the raw materials used. Humans have historically sought ways to turn raw materials, such as ore, wood, and foodstuffs, into finished products, such as metal goods furniture and processed foods. By refining and processing this raw material into something more useful, individuals and businesses have added value. This added value increased the price of finished products, rendering manufacturing a profitable endeavor. People began to specialize in the skills required to manufacture goods while others provided funds to businesses to purchase tools and materials.

How products are manufactured has changed over time. The amount and type of labor required in manufacturing vary according to the type of product being produced. On one end of the spectrum, products are manufactured by hand or through the use of basic tools using more traditional processes. This type of manufacturing is associated with decorative art, textile or leather work, carpentry and some metal work. At the other end of the spectrum, mechanization is used to produce items on a more industrial scale. This type of manufacturing does not require as much manual manipulation of materials and is often associated with mass production.

The industrial process used to turn raw materials into products in high volumes emerged during the Industrial Revolution of the 19th century. Before this period, handmade products dominated the market. The development of steam engines and related technologies allowed companies to use machines in the manufacturing process reducing the number of personnel required to produce goods while also increasing the volume of goods that could be produced.

Mass production and assembly line manufacturing allowed companies to create parts that could be used interchangeably and allowing finished products to be made more readily by reducing the need for part customization. The use of mass production techniques in manufacturing was popularized by the Ford Motor Company in the early 20th century. Computers and precision electronic equipment have since allowed companies to pioneer high-tech manufacturing methods. Products made using these methods typically carry a higher price but also require more specialized labor and more expensive capital inputs.

The skills required to operate machines and develop the processes used in manufacturing have changed drastically over time. Many low skill manufacturing jobs have shifted from developed countries to developing countries because labor in developing countries tends to be less expensive. More skilled manufacturing, particularly of precision and high-end products, tends to be undertaken in developed economies. Technology has made manufacturing more efficient and employees more productive; therefore, although the volume and number of goods manufactured have increased, the number of workers required has declined.

Economists and government statisticians use various ratios when evaluating the role manufacturing plays in the economy. Manufacturing value added (MVA), for example, is an indicator that compares manufacturing output to the size of the overall economy. It is expressed as a percentage of GDP - gross domestic product. The ISM Manufacturing Index uses surveys of manufacturing firms to estimate employment, inventories and new orders and is an indicator of the health of the manufacturing sector.

## Types of Manufacturing Processes

Manufacturing is a very simple business, the owner buys the raw material or component parts to manufacture a finished product. To function as a business the manufacturer needs to cover costs, meet demand and make a product to supply the market.

A factory operates one of three types of manufacturing production:

- Make-To-Stock (MTS) – A factory produces goods to stock stores and showrooms. By predicting the market for their goods, the manufacturer will plan production activity in advance. If they produce too much they may need to sell surplus at a loss and in producing too little they may miss the market and not sell enough to cover costs.

- Make-To-Order (MTO) – The producer waits for orders before manufacturing stock. Inventory is easier to control and the owner does not need to rely as much on market demand. Customer waiting time is longer though and the manufacturer needs a constant stream of orders to keep the factory in production.

- Make-To-Assemble (MTA) – The factory produces component parts in anticipation of orders for assembly. By doing this, the manufacturer is ready to fulfill customer orders but if orders do not materialize, the producer will have a stock of unwanted parts.

## Keeping Risks Under Control is Key

With all three types of manufacturing there are risks. Supply too much and you flood the market, causing a drop in price and a drop in profits. By not meeting demand, the customer may go elsewhere with a drop in sales for the manufacturer. Quality control is also a big factor in successful manufacturing. The manufacturer will need to keep a close eye on quality of product from beginning to end, with many tests along the way. If mistakes happen, the long-term consequences may be serious.

A manufacturing business may need many parts for the complicated assembly of a quality product or just the few for making a simple good. Keeping production costs to a minimum, having good quality control and excellent sales management are key to reducing the risk in any type of manufacturing.

# Computer Integrated Manufacturing

Computer-integrated manufacturing (CIM) is the use of computer techniques to integrate manufacturing activities. These activities encompass all functions necessary to translate customer needs into a final product. CIM starts with the development of a product concept that may exist in the marketing organization; includes product design and specification, usually the responsibility of an engineering organization; and extends through production into delivery and after-sales activities that reside in a field service or sales organization. Integration of these activities requires that accurate information be available when needed and in the format required by the person or group requesting the data. Data may come directly from the originating source or through an intermediate database according to Jorgensen and Krause. CIM systems have emerged as a result of the developments in manufacturing and computer technology.

The various processes involved in a CIM are listed as follows:

Stage One - Computer Aided Design. A product is designed totally on computer. When complete it is tested or its functions simulated on screen before even a prototype is made. If a circuit is involved it is designed by using software and tested on screen. Improvements/alterations are made to the design using the same CAD software.

Stage Two - Prototype Manufacture. Prototypes are manufactured on machines such as 3D printers which produce an accurate 3D model. CNC routers and laser cutters may also be used to produce a realistic model. Sometimes working models are manufactured.

Stage Three - The computer system controlling the plant works out the most efficient method of manufacture. It calculates costs, production methods, numbers to be manufactured, storage and distribution.

THE COMPUTER SYSTEM CONTROLS EVERY ASPECT
FROM DESIGN TO MANUFACTURE TO STORAGE AND DISTRIBUTION

Stage One - Computer aided design.

Stage Two - Prototype Manufacture

Stage Three - Costs and production methods calculated.

Stage Four - Materials automatically ordered.

Stage Five - Manufacturing begins using CAM

Stage Six - Quality control is applied at every stage

NO

YES

Stage Seven - The product is assembled by robots

Stage Six - Quality control is applied at every stage

NO

YES

Stage Nine - Product distribution.

Stage Ten - Financial accounts are updated

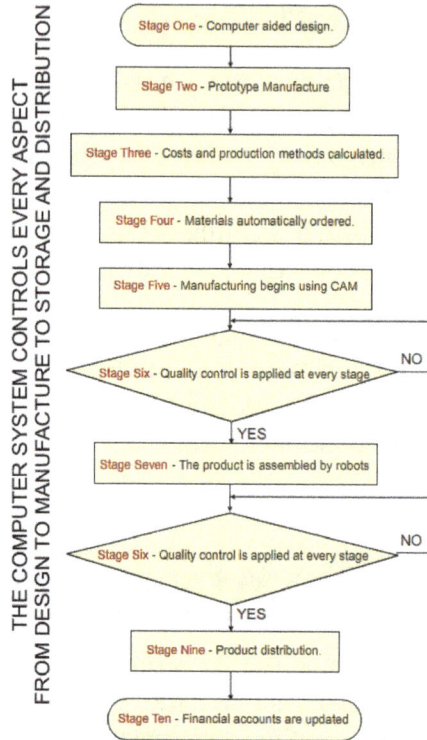

**Stage Four** - The computer system orders the necessary materials to manufacture the product. Keeping costs to a minimum. The 'just in time' philosophy is applied. This means that materials / components are ordered as needed. Very little is stored at the factory. Usually only enough materials are stored to keep the factory going for a small number of days. Materials are automatically reordered when required, to keep the factory working smoothly and continuously.

**Stage Five** - Manufacturing begins with the product being made using CAM (Computer Aided Manufacture). Computers control CNC machines such as laser cutters, CNC routers and CNC lathes.

**Stage Six** - Quality control is applied at every stage. The product is tested using computer control inspections. For instance, the accuracy of manufacture is tested automatically. This ensures that the product is manufactured to the correct sizes.

**Stage Seven** - The product is assembled by robots. This is automated (controlled) by the computer system.

**Stage Eight** - The product is quality checked before being stored for distribution to the customer. All storage is automated. This means that computer controlled vehicles move the finished product from the manufacturing area to storage. The computer systems keep track of every individual product. Products are bar coded which are constantly scanned and recorded by the computer system.

Stage Nine - The product is automatically moved from store to awaiting lorries/trucks for distribution to the customer.Stage Ten - Financial accounts are updated, bills chased up and paid by the computer system.

The term computer-integrated manufacturing was coined by Dr. Joseph Harrington in his 1974 book bearing that name. Until the 1970s, the most aggressive and success-ful automation was seen in production operations. Discrete parts manufacturing used highly mechanized machines that were driven and controlled by cams and complex devices such as automatic screw machines. Process manufacturers made use of these cam-driven controllers and limit switches for operations such as heat treating, filling and canning, bottling, and weaving states Robert Thacker of the Society of Manufac-turing Engineers. The historical approach to automation focused on individual activi-ties that result in the incorporation of large amounts of computerized activities. In the 1980s, managing information became an important issue.

## Benefits

According to the U.S. National Research Council, CIM improves production produc-tivity by 40 to 70 percent, as well as enhances engineering productivity and quality. CIM can also decrease design costs by 15 to 30 percent, reduce overall lead time by 20 to 60 percent, and cut work-in-process inventory by 30 to 60 percent. Managers who use CIM believe that there is a direct relationship between the efficiency of information management and the efficiency and the overall effectiveness of the manufacturing en-terprise. Thacker's view is that many CIM programs focus attention on the efficiency of information management and the problems that come with it instead of developing new and more sophisticated manufacturing machines, material transformation pro-cesses, manufacturing management processes, and production facilities. Computer-in-tegrated manufacturing can be applied to nonmanufacturing organizations by changing the manufacturing focus toward a service orientation. CIM and Job Definition Format (JDF) are becoming increasingly beneficial to printing companies to streamline their production process.

## Plan

A plan for a CIM system should provide a description of projects for automating activ-ities, assisting activities with technology, and integrating the information flows among these activities. The planning process includes six crucial steps:

- Project activation
- Business assessment
- Business modeling
- Needs analysis

- Conceptual design

- CIM plan consolidation and economic analysis

This process, according to Jorgensen and Krause, also acts as a building block for the future of the organization integrating these functions in order to diminish them as an impediment to integration.

## Conceptual Design

The conceptual design of a CIM environment consists of individual systems that fulfill the required capabilities, an overall architecture incorporating the systems and the communication links, and a migration path from the current systems architecture. Functional requirements must be compared to the current inventory of systems and available technology to determine system availability. Jorgensen and Krause state that the following techniques are used in satisfying system requirements:

- Exploiting unused and available functional capabilities of current systems;

- Identifying functional capabilities available for, but not installed on, current in-house systems;

- locating systems that are commercially available but not currently in-house;

- Recognizing state-of-the-art technology that is not immediately commercially available on a system;

- Foreseeing functional capabilities of systems on the technical horizon;

- Determining whether the requirement is beyond the capabilities of systems on the technical horizon.

## Managing a CIM

Managers must understand that short-term goals must support the long-term goal of implementing a CIM. Top management establishes long-term goals for the company and envisions the general direction of the company. The middle management then creates objectives to achieve this goal. Upper management sees the focus as being very broad, whereas middle management must have a more narrow focus.

In deciding to implement a CIM, there are three perspectives that must be considered: the conceptual plan, the logical plan, and the physical plan. The conceptual plan is used to demonstrate a knowledgeable understanding of the elements of CIM and how they are related and managed. Thacker goes on to say that the conceptual plan states that by integrating the elements of a business, a manager will produce results better and faster than those same elements working independently.

The logical plan organizes the functional elements and logically demonstrates the relationships and dependencies between the elements. Thacker details that it further shows how to plan and control the business, how to develop and connect an application, communications, and database network.

The physical plan contains the actual requirements for setting the CIM system in place. These requirements can include equipment such as hardware, software, and work cells. The plan is a layout of where the computers, work stations, robots, applications, and databases are located in order to optimize their use within the CIM and within the company. According to Thacker, sooner or later it becomes the CIM implementation plan for the enterprise.

CIM is challenged by technical and cultural boundaries. The technical challenge is first complicated by the varying applications involved. Thacker claims that it is also complicated by the number of vendors that the CIM serves as well as incompatibility problems among systems and lack of standards for data storage, formatting, and communications. Companies must also have people who are well-trained in the various aspects of CIM. They must be able to understand the applications, technology, and communications and integration requirements of the technology.

CIM cultural problems begin within the division of functional units within the company such as engineering design, manufacturing engineering, process planning, marketing, finance, operations, information systems, materials control, field service, distribution, quality, and production planning. CIM requires these functional units to act as whole and not separate entities. The planning process represents a significant commitment by the company implementing it. Although the costs of implementing the environment are substantial, the benefits once the system is in place greatly outweigh the costs. The implementation process should ensure that there is a common goal and a common understanding of the company's objectives and that the priority functions are being accomplished by all areas of the company according to Jorgensen and Krause.

## Computer Aided Technologies

The major objectives of computer-aided technology (CAx) are to simplify and to improve human's work (engineer, architect, physician, surgeon, etc.), by using the computer as an indispensable tool to solve a problem in a certain field (engineering and production, medicine, architecture, business, teaching, economy, etc.) .

The advanced computer-aided technologies (CAx) are focused on solving specific problems by increasing human's creativity and innovation obtained through collecting, using, and sharing information between interdisciplinary teams.

Computer-aided technologies in X field are general terms to define a technology, from a specific field of work, which is computed-aided. The substitute for X includes engineering (CAx-E), medicine (CAx-M), natural science (CAx-S), education (CAx-Ed), etc.

Nowadays, computer-aided technologies are not islands of automation, being integrated in general context of Lifecycle Management in X field. The concepts used to define the lifecycle management in engineering and medical field.

- Product lifecycle management (PLM) in the industry field

- Human lifecycle management (HUM) or health management across the human lifecycle.

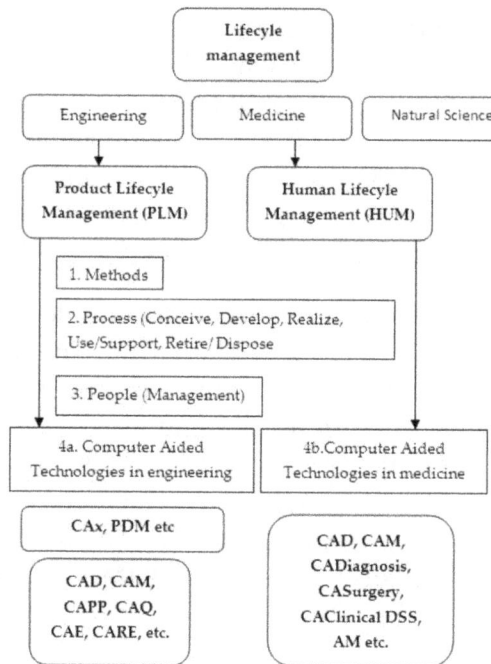

## Computer-aided Technologies in Product Lifecycle Management

Product lifecycle management (PLM) is the process of managing the entire lifecycle of a product including conception, design, manufacturing, quality control, use, service, and disposal of products, having integrated people, data, methods, CAx tools, processes, and business systems. PLM is a digital paradigm, products being managed with digital computer, digital information, and digital communication .

The main benefits of product lifecycle management (PLM) for the industry field are faster time to-market, improved productivity, better product quality, decreased cost of new product introduction, improved design review, and approval processes, identifying potential sales opportunities and revenue contributions and reducing environmental impacts at end-of-life.

PLM emerged from tools such as CAD, CAM, and PDM, being viewed as the integration of these tools with innovative technologies (e.g., additive manufacturing, reverse engineering), methods, people, and the processes through all stages of a product's life .

To improve the product lifecycle management, a lot of methods and techniques are used such as concurrent engineering, bottom-up design, top-down design, both-ends-against-the middle design, design in context, design for X (DFX), TRIZ, lean production, design for six sigma (DFSS), total quality management (TQM), and failure mode effects analysis (FMEA).

Concurrent engineering or simultaneous engineering is a workflow that, instead of working sequentially through stages, carries out a number of tasks in parallel.

The most used design for X method is design for manufacture and assembly (DFMA) that is the combination of two methodologies: design for manufacture (DFM) and design for assembly (DFA), which mean the design of the parts for ease of manufacturing and the design of the product for ease of assembly.

Product data management (PDM) is the business function often within product lifecycle management (PLM) that is responsible for the management and publication of product data.

The tools used to access the information and knowledge regarding the product data are the computer-aided technologies (CAx). Computer-aided technologies (CAx) are the use of computer technology to aid in the design, analysis, production planning, manufacture of products, etc.

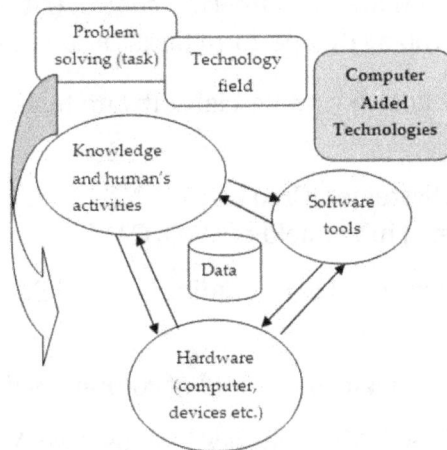

A CAx system can work like an "island of automation" or it can be integrated in the PLM system, interacting with other "islands of automation". Thus, the advanced CAx tools merge many different aspects of the product lifecycle management, including design, manufacturing, etc. CAx can be integrated, also, with other computational systems of management and planning of trials and output, such as MRP (material resource planning), ERP (enterprise resource planning), EDM (electronic document management), and PDM (product data management).

A CAx system may be defined, in generally, having the following main components:

- Hardware component consisting in computer and interactive devices
- Software packages
- Data
- Knowledge and human's activities.

Computer-aided design (CAD), computer-aided engineering (CAE), computer-aided manufacturing (CAM), computer-aided process planning (CAPP), and computer-aided quality assurance (CAQA) are the most known and mature computer-aided technologies.

Computer-aided design is the computer-aided technology that involves the computer to assist in the creation, modification, analysis, and optimization of a design and design documentation.

Computer-aided engineering (CAE) is the computer-aided technology that involves the computer to analyze, simulate, and optimize the CAD geometry. CAE tools are available for a wide range of analyses: stress analysis, deformation, heat transfer, fluid flow, magnetic field distribution, kinematics, and dynamic analysis, etc.

Computer-aided manufacturing (CAM) is the computer-aided technology that involves the computer in planning, control, and management of manufacturing of any product. The most mature areas of CAM are the numerical control (NC) of the machine tools and programming of industrial robots that perform tasks as assembly, welding, etc.

The following are the most known commercial software tools for computer-aided technologies:

- CATIA by Dassault Systemes, Creo by PTC, NX by Siemens, Power Shape/ Power Mill by Delcam etc., in the field of CAD/CAM
- Material is Magic's and Netfabb Studio, in the field of 3D-printing/ additive manufacturing
- Rapid Form and Geo magic in the field of computer-aided reverse engineering
- Ansys, Abaqus, COMSOL Multi physics, Adams, LMS Virtual. Lab are focused on CAE.

The computer-aided technology tools used in the engineering field are:

- CAx in engineering field X=design, analysis, process planning, manufacturing, quality, innovation, etc.;

- CAD - Computer-aided design;

- CAM- Computer-aided manufacturing . Software to control machine tools and related machinery in the manufacturing of work pieces;

- CAPP- Computer-aided process planning system is the bridge between CAD and CAM;

- CAE -Computer-aided engineering. Software to aid the simulation of mechanical, strength, temperature, pressure, etc;

- FEA- Finite element analysis is the practical application of the finite element method (FEM), which involves the use of numerical methods in structure analysis, dynamics, thermal analysis, etc;

- CFD- Computational fluid dynamics uses numerical analysis and algorithms to analyze problems that involve fluid flows;

- MBDS –Multi body dynamics simulations;

- CAQ -Computer-aided quality;

- CAQA -Computer-aided quality assurance;

- CAInsp -Computer-aided inspection;

- CAI- Computer-aided innovation  is an emerging domain in the array of computer-aided;

- CARE -Computer-aided reverse engineering  has the aim to capture the geometry of an existing physical model, through digitization, and to create a 3D virtual model that is used then in different applications;

- CATAM - Computer-aided technologies for additive manufacturing. Supporting the design, simulation and process planning for additive manufacturing;

- RapidX - RapidX  is a generic term for rapid technologies, e.g., rapid prototyping (RP), rapid tooling;

- (RT), and rapid manufacturing (RM);

- CAD Composite  -Computer-aided design in composite material technology;

- CAM Composite-Computer-assisted manufactured composite;

- CAMT- Computer-aided manufacturing technologies;

- CAPE -Computer-aided production engineering;

- CA Maintenance-Computer-aided maintenance;

- CAMSE -Computer-aided manufacturing system Engineering. CAMSE is defined as the use of computerized tools in the application of scientific and engineering methods to the problem of the design and implementation of manufacturing systems;

- CAT- Computer-aided techniques for tolerance analysis computer aided tolerance;

- CAAD - Computer-aided architectural design;

- CAID - Computer-aided industrial design;

- CAW- Computer-aided welding;

- CAWFD - Computer-aided welding fixture design;

- CAFD - Computer-aided fixture design;

- CAMAP - Computer-aided mechanical assembly planning.

- CADM - Computer- aided design and manufacture.

- CAD/CAM/CAQ- Integrated CAD/CAM/CAQ system.

- CIM- Computer integrated manufacturing  has the purpose to tying "the separate islands of automation" together, including the computer-aided  design, computer-aided planning, computer- aided manufacturing, and computer-aided quality assurance into an efficient system.

## Computer Aided Manufacturing

Computer Aided Manufacturing (CAM) is the use of software and computer-controlled machinery to automate a manufacturing process.

Based on that definition, you need three components for a CAM system to function:

- Software that tells a machine how to make a product by generating tool paths.

- Machinery that can turn raw material into a finished product.

- Post Processing that converts tool paths into a language machines can understand.

These three components are glued together with tons of human labor and skill. As an industry we've spent years building and refining the best manufacturing machinery around. Today, there's no design too tough for any capable machinist shop to handle.

### Relationship between CAM, CAD, and BIM

CAM tends to go hand-in-hand with computer aided design (CAD) and building information modeling (BIM), at least as far as its application in the construction industry goes. CAD allows architects and members of the design team to make drawings in 2D or create entire 3D models using computer software. This has a number of advantages over traditional pen and paper drawings, including the ability to redraw and redesign easily, to save component parts in databases and (in the case of 3D CAD) the ability to rotate and fly into or through the model.

BIM utilizes CAD but allows for collaboration between different design and construction stakeholders, who can work on their own models while accessing and combining with other parties' models to create a central 'federated' BIM model. Additional data relating to elements such as cost and time can also be added.

The data from CAD and BIM drawings and models can be extracted and used to create the G Code used in computer aided manufacturing. That closes the gap existing between the design and manufacturing stages and allows for the accurate realization of drawings, models and designs.

## Usage of CAM in the Construction Industry

CAM is being used onsite all around the world, although as of yet, it is still far from commonplace. CAM generally falls into two broad types: reductive and additive.

Reductive processes involve getting rid of material, and this includes the previous example of guiding a cutting tool to cut out a section of cladding. These cutting and shaping processes are currently the more commonly used types of CAM, and the laser cutting of sheet metal is certainly becoming more common. CNC (computer numerical control) routing uses a spinning component to carve materials into the desired shape while laser and water cutting can be used on relatively thin panels and pieces.

Additive processes involve adding material. They are far less common at present, but the arrival of 3D printers makes this a very exciting area. We could see walls and whole structures being 'printed', while robotics open up another avenue. Robot bricklayers and saws have already been trialed, and in some cases, deployed on construction sites.

Modular construction is another area where the potential for CAM is huge. In this method, buildings and other structures are assembled from components that are prefabricated offsite in manufacturing plants before being transported to the construction site for assembly. Sweden is a world leader in modular construction, with 84% of detached homes in the country using some prefabricated elements. Modular construction is also taking off in Germany and, while it's not quite as popular in the UK and the USA, advances in CAM technology can be used to greatly enhance the efficiency of offsite modular building, speeding up and improving the accuracy of the component construction.

One example of modular building is GSK's 'factory in a box'. Created using CAD and BIM systems, this provides a colour-coded pharmaceutical factory that that can be shipped to developing areas in crates and put together like an altogether more impressive set of flat-pack furniture.

## Benefits of CAM

The benefits of CAM include a properly defined manufacturing plan that delivers expected results in production:

- CAM systems can maximize utilization of a full range of production equipment, including high speed, 5-axis, multi-function and turning machines, electrical discharge machining (EDM) and CMM inspection equipment.

- CAM systems can aid in creating, verifying, and optimizing NC programs for optimum machining productivity, as well as automate the creation of shop documentation.

- Advanced CAM systems with product lifecycle management (PLM) integration can provide manufacturing planning and production personnel with data and process management to ensure use of correct data and standard resources.

- CAM and PLM systems can be integrated with DNC systems for delivery and management of files to CNC machines on the shop floor.

There are some limitations. CAM-enabled machines are generally designed for a particular task and are not incredibly versatile, although new systems and designs are emerging all the time.

They also need an upfront investment and skilled operators and programmers. Once in place, however, they could potentially bring large savings in time and efficiency, thereby reducing costs and saving companies thousands.

## Digital Manufacturing

Digital manufacturing is the use of an integrated, computer-based system comprised of simulation, 3D visualization, analytics and collaboration tools to create product and manufacturing process definitions simultaneously. Digital manufacturing evolved from manufacturing initiatives such as design for manufacturability (DFM), computer-integrated manufacturing (CIM), flexible manufacturing and lean manufacturing that highlight the need for collaborative product and process design.

Many of the long-term benefits from product lifecycle management (PLM) cannot be achieved without a comprehensive digital manufacturing strategy. Digital manufacturing is a key point of integration between PLM and shop floor applications and equipment, enabling the exchange of product-related information between design and manufacturing groups. This alignment allows manufacturing companies to achieve time-to-market and volume goals, as well as realize cost savings by reducing expensive downstream changes.

## Digital Manufacturing Connects Processes

Digital manufacturing requires integration between PLM, ERP, shop floor applications and equipment to enable the exchange of product-related information between digital design and physical manufacturing execution. Manufacturers can achieve time-to-market and volume goals, as well as realize cost savings by establishing a digital thread that provides a formal framework for digital exchange that can analyze data throughout the product lifecycle, transforming it into actionable information. The digital thread integrates as-designed requirements, validation and inspection records, as-built data, as-flown data, and as-maintained data.

Smart, connected products can send customer data to product managers to help anticipate demand and meet ongoing maintenance needs. The result is better designed products maintained in detail throughout the product life cycle. By involving customers during the entire product lifecycle, requirements can be fulfilled faster and with far fewer iterations.

## Three Dimensions of Digital Manufacturing

Digital manufacturing can be broken down into three dimensions:

(a)  Product Life Cycle,

(b)  Smart Factory,

(c)  Value Chain Management.

The Product Life Cycle starts with an engineering design definition and follows through sourcing, production and service life. Digital data for each step includes every incorporated revision, any approved deviations from design specifications and how these are executed across the lifecycle.

The Smart Factory is all about automation. It encompasses smart machines, sensors and tooling to provide workers with real-time data about the processes they are executing. It forms the bridge between Operations Technology (OT) that exchanges data directly with machines and tooling, and Information Technology (IT) systems and apps. Both are enhanced by business intelligence systems that perform in-depth analysis. This leads to real-time visibility of factory processes, process control optimization, and insight into potential areas of performance or process improvement.

Value Chain Management focuses on minimizing resources and accessing value at each stakeholder function along the chain. It results in optimal process integration, decreased inventories, better products, and enhanced customer satisfaction.

## Benefits of Digital Manufacturing

Digital manufacturing brings together complex manufacturing processes across departments, and eliminates paper processes that can be fraught with errors and repeated information. Benefits include:

- Increased efficiency through automated exchange of data
- Avoidance of costly errors due to missed or misinterpreted data
- Quicker turnaround at all levels of the value chain
- Greater insight at critical decision points
- Real-time visibility into the effects of changes to processes, equipment, systems or components
- Faster pace of innovation
- Lowered cost of production and maintenance.

## Tooling and Processes

There are many different tooling processes that digital manufacturing utilizes. However, every digital manufacturing process involves the use of computerized numerical controlled machines (CNC). This technology is crucial in digital manufacturing as it not only enables mass production and flexibility, but it also provides a link between a CAD model and production. The two primary categories of CNC tooling are additive and subtractive. Major strides in additive manufacturing have come about recently and are at the forefront of digital manufacturing. These processes allow machines to address every element of a part no matter the complexity of its shape.

## Examples of Additive Tooling And Processes

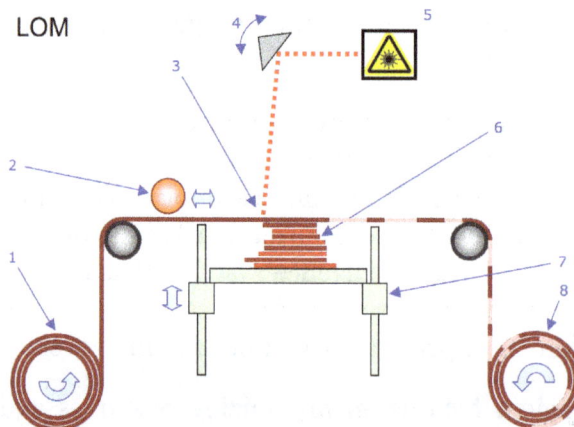

Example of Laminated object manufacturing process Laminated object manufacturing: principle drawing Supply roll

Stereolithography - In this process, solid parts are formed by solidifying layers of a photopolymer with ultraviolet light. There is a wide range of acrylics and epoxies that are used in this process.

- Ink-Jet Processing - Although the most widely used ink-jet process is used for printing on paper, there are many that are applied in engineering. This process involves a print head depositing layers of liquid material onto a filler powder in the shape of the desired object. After the powder is saturated, a fresh new layer of powder is added continually until the object is built.

- Laser sintering and fusion - This process utilizes heat produced by infrared lasers to bond a powdered material together to form a solid shape.

- Solid Ground Curing - A layer of liquid photopolymer is spread over a platform. An optical mask is generated and laid over the polymer. A UV lamp cures the resin that is not blocked by the mask. Any remaining liquid is removed and the voids are filled with wax. Liquid resin is spread over the layer that was just produced and the process is repeated. When the part is finished, the wax can be melted out of the voids.

- Laminated-Object Manufacturing - A sheet material is laid on a platform and a laser cuts the desired contour. The platform is lowered by one sheet thickness and a new sheet is laid with a layer of thermal adhesive between the two sheets. A heated roller presses the sheets together and activates the adhesive. The laser cuts the contours of this layer and the process is repeated. When the part is finished, the leftover sheet material around the perimeter of the part must be removed. The final part is coated with sealant.

- Fused filament fabrication- FFF is the most commonly used form of 3-D printing. Thermoplastic material is heated just beyond solidification and extruded onto a platform in the desired shape. The platform is lowered, and the next layer is extruded onto the previous layer. The process is repeated until the part is complete.

## Examples of Subtractive Tooling and Processes

A CNC water jet cutter is an example of the types of computer controlled tooling that are essential to digital manufacturing.

- Water Jet Cutting - A water jet cutter is a CNC tool that uses a high pressure stream of water, often mixed with an abrasive material, to cut shapes or patterns out of many types of materials.

- Milling - A CNC mill uses a rotational cutting tool to remove material from a piece of stock. Milling can be performed on most metals, many plastics, and all types of wood.

- Lathe - A CNC lathe removes material by rotating the work-piece while a stationary cutting tool is brought into contact with the material.

# Control Theory

Control theory is a branch of mathematics and engineering, which defines the conditions needed for a system to maintain a controlled output in the face of input variation.

Although control theory has deep connections with classical areas of mathematics, such as the classical areas of mathematics, calculus of variations and the theory of differential equations, it did not become a field in its own right until the late 1950s and early 1960s. At that time, problems arising in engineering and economics were recognized as variants of problems in differential equations and in the calculus of variations, though they were not covered by existing theories. At first, special modifications of classical techniques and theories were devised to solve individual problems. It was then recognized that these seemingly diverse problems all had the same mathematical structure, and control theory emerged.

As long as human culture has existed, control has meant some kind of power over the environment. For example, cuneiform fragments suggest that the control of irrigation systems in Mesopotamia was a well-developed art at least by the 20th century BC. There were some ingenious control devices in the Greco-Roman culture, the details of which have been preserved. Methods for the automatic operation of windmills go back at least to the European Middle Ages. Large-scale implementation of the idea of control, however, was impossible without a high level of technological sophistication, and the principles of modern control started evolving only in the 19th century, concurrently with the Industrial Revolution. A serious scientific study of this field began only after World War II.

Although control is sometimes equated with the notion of feedback control (which involves the transmission and return of information)—an isolated engineering invention, not a scientific discipline—modern usage favours a wider meaning for the term. For instance, control theory would include the control and regulation of machines, muscular coordination and metabolism in biological organisms, and design of prosthetic devices, as well as broad aspects of coordinated activity in the social sphere such as optimization of business operations, control of economic activity by government policies, and even control of political decisions by democratic processes. If physics is the science of understanding the physical environment, then control theory may be viewed as the science of modifying that environment, in the physical, biological, or even social sense.

Much more than even physics, control is a mathematically oriented science. Control principles are always expressed in mathematical form and are potentially applicable to any concrete situation. At the same time, it must be emphasized that success in the use of the abstract principles of control depends in roughly equal measure on basic scientific knowledge in the specific field of application, be it engineering, physics, astronomy, biology, medicine, econometrics, or any of the social sciences.

## Examples of Modern Control Systems

To clarify the critical distinction between control principles and their embodiment in a real machine or system, the following common examples of control may be helpful.

## Machines That Cannot Function without Control

Many basic devices must be manufactured in such a way that their behaviour can be modified by means of some external control. Generally, the same effect cannot be brought about (in practice and sometimes even in theory) by any intrinsic modification of the characteristics of the device. For example, transistor amplifiers introduce intolerable distortion in sound systems when used alone, but properly modified by a feedback control system they can achieve any desired degree of fidelity. Another example involves powered flight. Early pioneers failed, not because of their ignorance of the laws of aerodynamics but because they did not realize the need for control and were unaware of the basic principles of stabilizing an inherently unstable device by means of control. Jet aircraft cannot be operated without automatic control to aid the pilot, and control is equally critical for helicopters. The accuracy of inertial navigation equipment cannot be improved indefinitely because of basic mechanical limitations, but these limitations can be reduced by several orders of magnitude by computer-directed statistical filtering, which is a variant of feedback control.

## Control of Machines

In many cases, the operation of a machine to perform a task can be directed by a human (manual control), but it may be much more convenient to connect the machine directly to the measuring instrument (automatic control); e.g., a thermostat may be used to turn on or off a refrigerator, oven, air-conditioning unit, or heating system. The dimming of automobile headlights, the setting of the diaphragm of a camera, and the correct exposure for colour prints may be accomplished automatically by connecting a photocell directly to the machine in question. Related examples are the remote control of position (servomechanisms) and speed control of motors (governors). It is emphasized that in such cases a machine could function by itself, but a more useful system is obtained by letting the measuring device communicate with the machine in either a feed forward or feedback fashion.

## Control of Large Systems

More advanced and more critical applications of control concern large and complex systems the very existence of which depends on coordinated operation using numerous individual control devices (usually directed by a computer). The launch of a spaceship, the 24-hour operation of a power plant, oil refinery, or chemical factory, and air traffic control near a large airport are examples. An essential aspect of these systems is that human participation in the control task, although theoretically possible, would be wholly impractical; it is the feasibility of applying automatic control that has given birth to these systems.

## Biocontrol

The advancement of technology (artificial biology) and the deeper understanding of the processes of biology (natural technology) has given reason to hope that the two can be combined; man-made devices should be substituted for some natural functions. Examples are the artificial heart or kidney, nerve-controlled prosthetics, and control of brain functions by external electrical stimuli. Although definitely no longer in the science-fiction stage, progress in solving such problems has been slow not only because of the need for highly advanced technology but also because of the lack of fundamental knowledge about the details of control principles employed in the biological world.

## Robots

On the most advanced level, the task of control science is the creation of robots. This is a collective term for devices exhibiting animal-like purposeful behaviour under the general command of (but without direct help from) humans. Highly specialized industrial manufacturing robots are already common, but real breakthroughs will require fundamental scientific advances with regard to problems related to pattern recognition and thought processes.

# Principles of Control

The scientific formulation of a control problem must be based on two kinds of information:

- The behaviour of the system must be described in a mathematically precise way.

- The purpose of control (criterion) and the environment (disturbances) must be specified, again in a mathematically precise way.

Information of type A means that the effect of any potential control action applied to the system is precisely known under all possible environmental circumstances. The choice of one or a few appropriate control actions, among the many possibilities that may be available, is then based on information of type B. This choice is called optimization.

The task of control theory is to study the mathematical quantification of these two basic problems and then to deduce applied mathematical methods whereby a concrete answer to optimization can be obtained. Control theory does not deal directly with physical reality but with mathematical models. Thus, the limitations of the theory depend only on the agreement between available models and the actual behaviour of the system to be controlled. Similar comments can be made about the mathematical representation of the criteria and disturbances.

Once the appropriate control action has been deduced by mathematical methods from the information mentioned above, the implementation of control becomes a technological task, which is best treated under the various specialized fields of engineering. The detailed manner in which a chemical plant is controlled may be quite different from that of an automobile factory, but the essential principles will be the same. Hence further discussion of the solution of the control problem will be limited here to the mathematical level.

To obtain a solution in this sense, it is convenient to describe the system to be controlled, which is called the plant, in terms of its internal dynamical state. By this is meant a list of numbers (called the state vector) that expresses in quantitative form the effect of all external influences on the plant before the present moment, so that the future evolution of the plant can be exactly given from the knowledge of the present state and the future inputs. This situation implies that the control action at a given time can be specified as some function of the state at that time. Such a function of the state, which determines the control action that is to be taken at any instant, is called a control law. This is a more general concept than the earlier idea of feedback; in fact, a control law can incorporate both the feedback and feed forward methods of control.

In developing models to represent the control problem, it is unrealistic to assume that every component of the state vector can be measured exactly and instantaneously. Consequently, in most cases the control problem has to be broadened to include the further problem of state determination, which may be viewed as the central task in statistical prediction and filtering theory. In principle, any control problem can be solved in two steps:

(1) Building an optimal filter (a so-called Kalman filter) to determine the best estimate of the present state vector.

(2) Determining an optimal control law and mechanizing it by substituting into it the estimate of the state vector obtained in step 1.

In practice, the two steps are implemented by a single unit of hardware, called the controller, which may be viewed as a special-purpose computer. The theoretical formulation given here can be shown to include all other previous methods as a special case; the only difference is in the engineering details of the controller.

The mathematical solution of a control problem may not always exist. The determination of rigorous existence conditions, beginning in the late 1950s, has had an important effect on the evolution of modern control, equally from the theoretical and the applied point of view. Most important is controllability; it expresses the fact that some kind of control is possible. If this condition is satisfied, methods of optimization can pick out the right kind of control using information of type B.

The controllability condition is of great practical and philosophical importance. Because the state-vector equations accurately represent most physical systems, which only have small deviations about their steady-state behaviour, it follows that in the natural world small-scale control is almost always possible, at least in principle. This fact of nature is the theoretical basis of practically all the presently existing control technology. On the other hand, little is known about the ultimate limitations of control when the models in question are not linear, in which case small changes in input can result in large deviations. In particular, it is not known under what conditions control is possible in the large, that is, for arbitrary deviations from existing conditions. This lack of scientific knowledge should be kept in mind in assessing often-exaggerated claims by economists and sociologists in regard to a possible improvement in human society by governmental control.

## References

- Mourtzis, Dimitris (2015). "The role of simulation in digital manufacturing: applications and outlook". International Journal of Computer Integrated Manufacturing

- Computer-integrated-manufacturing-cim-30965: techopedia.com, Retrieved 18 July 2018

- Computer-aided-manufacturing-beginners, fusion-360: autodesk.com, Retrieved 11 May 2018

- What-is-cam-computer-aided-manufacturing, construction-industry, Retrieved 28 June 2018

- Digital-manufacturing-13157: plm.automation.siemens.com, Retrieved 10 April 2018

- What-is-digital-manufacturing: ibaset.com, Retrieved 21 May 2018

- Control-theory: nature.com, Retrieved 28 May 2018

- Control-theory-mathematics, science: britannica.com, Retrieved 31 March 2018

- Huang, Samuel (July 2013). "Additive manufacturing and its societal impact: a literature review". International Journal of Advanced Manufacturing Technology

# 2

# Automation in Industrial Sector

Automation involves the use of control systems for the efficient execution of processes and operation of equipment and machinery with reduced or minimal human intervention. The aim of this chapter is to explore the need for automation in the industrial sector and its status in the modern scenario. It includes vital topics such as industrial automation, control loop, programmable logic controller and human machine interface.

## Automation

Automation is the application of machines to tasks once performed by human beings or, increasingly, to tasks that would otherwise be impossible. Although the term mechanization is often used to refer to the simple replacement of human labour by machines, automation generally implies the integration of machines into a self-governing system. Automation has revolutionized those areas in which it has been introduced, and there is scarcely an aspect of modern life that has been unaffected by it.

The term automation was coined in the automobile industry about 1946 to describe the increased use of automatic devices and controls in mechanized production lines. The origin of the word is attributed to D.S. Harder, an engineering manager at the Ford Motor Company at the time. The term is used widely in a manufacturing context, but it is also applied outside manufacturing in connection with a variety of systems in which there is a significant substitution of mechanical, electrical, or computerized action for human effort and intelligence.

In general usage, automation can be defined as a technology concerned with performing a process by means of programmed commands combined with automatic feedback control to ensure proper execution of the instructions. The resulting system is capable of operating without human intervention. The development of this technology has become increasingly dependent on the use of computers and computer-related technologies. Consequently, automated systems have become increasingly sophisticated and complex. Advanced systems represent a level of capability and performance that surpass in many ways the abilities of humans to accomplish the same activities.

Automation technology has matured to a point where a number of other technologies have developed from it and have achieved a recognition and status of their own. Robotics is one of these technologies; it is a specialized branch of automation in which the automated machine possesses certain anthropomorphic, or human-like, characteristics. The most typical humanlike characteristic of a modern industrial robot is its powered mechanical arm. The robot's arm can be programmed to move through a sequence of motions to perform useful tasks, such as loading and unloading parts at a production machine or making a sequence of spot-welds on the sheet-metal parts of an automobile body during assembly. As these examples suggest, industrial robots are typically used to replace human workers in factory operations.

## Principles and Theory of Automation

There are three basic building blocks of automation:

- A source of power to perform some action,

- Feedback controls,

- Machine programming. Almost without exception, an automated system will exhibit all these elements.

## Power Source

An automated system is designed to accomplish some useful action, and that action requires power. There are many sources of power available, but the most commonly used power in today's automated systems is electricity. Electrical power is the most versatile, because it can be readily generated from other sources (e.g., fossil fuel, hydroelectric, solar, and nuclear) and it can be readily converted into other types of power (e.g., mechanical, hydraulic, and pneumatic) to perform useful work. In addition, electrical energy can be stored in high-performance, long-life batteries.

The actions performed by automated systems are generally of two types:

(1) Processing;

(2) Transfer and positioning.

In the first case, energy is applied to accomplish some processing operation on some entity. The process may involve the shaping of metal, the molding of plastic, the switching of electrical signals in a communication system, or the processing of data in a computerized information system. All these actions entail the use of energy to transform the entity (e.g., the metal, plastic, electrical signals, or data) from one state or condition into another more valuable state or condition. The second type of action—transfer and positioning—is most readily seen in automated manufacturing systems designed to perform work on a product. In these cases, the product must generally be moved (transferred) from one location to another during the series of processing steps. At each processing location, accurate positioning of the product is generally required. In automated communications and information systems, the terms transfer and positioning refer to the movement of data (or electrical signals) among various processing units and the delivery of information to output terminals (printers, video display units, etc.) for interpretation and use by humans.

## Feedback Controls

Feedback controls are widely used in modern automated systems. A feedback control system consists of five basic components:

- Input,

- Process being controlled,

- Output,

- Sensing elements,

- Controller and actuating devices.

These five components are illustrated in the figure below. The term closed-loop feedback control is often used to describe this kind of system.

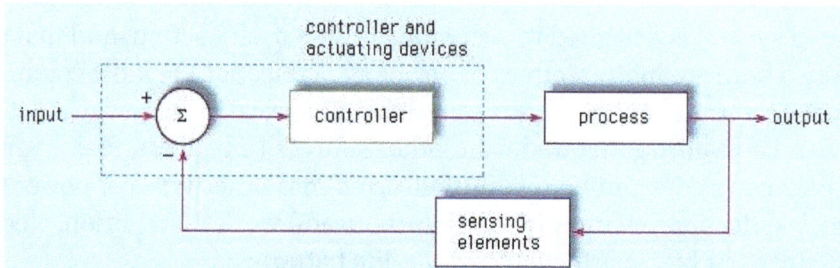

The components of a feedback control system and their relationships

The input to the system is the reference value, or set point, for the system output. This represents the desired operating value of the output. Using the previous example of the heating system as an illustration, the input is the desired temperature setting for a room. The process being controlled is the heater (e.g., furnace). In other feedback systems, the process might be a manufacturing operation, the rocket engines on a space shuttle, the automobile engine in cruise control, or any of a variety of other processes to which power is applied. The output is the variable of the process that is being measured and compared to the input; in the above example, it is room temperature.

The sensing elements are the measuring devices used in the feedback loop to monitor the value of the output variable. In the heating system example, this function is normally accomplished using a bimetallic strip. This device consists of two metal strips joined along their lengths. The two metals possess different thermal expansion coefficients; thus, when the temperature of the strip is raised, it flexes in direct proportion to the temperature change. As such, the bimetallic strip is capable of measuring temperature. There are many different kinds of sensors used in feedback control systems for automation.

The purpose of the controller and actuating devices in the feedback system is to compare the measured output value with the reference input value and to reduce the difference between them. In general, the controller and actuator of the system are the mechanisms by which changes in the process are accomplished to influence the output variable. These mechanisms are usually designed specifically for the system and consist of devices such as motors, valves, solenoid switches, piston cylinders, gears, power screws, pulley systems, chain drives, and other mechanical and electrical components. The switch connected to the bimetallic strip of the thermostat is the controller and actuating device for the heating system. When the output (room temperature) is below the set point, the switch turns on the heater. When the temperature exceeds the set point, the heat is turned off.

## Machine Programming

The programmed instructions determine the set of actions that is to be accomplished automatically by the system. The program specifies what the automated system should do and how its various components must function in order to accomplish the desired result. The content of the program varies considerably from one system to the next. In relatively simple systems, the program consists of a limited number of well-defined actions that are performed continuously and repeatedly in the proper sequence with no deviation from one cycle to the next. In more complex systems, the number of commands could be quite large, and the level of detail in each command could be significantly greater. In relatively sophisticated systems, the program provides for the sequence of actions to be altered in response to variations in raw materials or other operating conditions.

Programming commands are related to feedback control in an automated system in that the program establishes the sequence of values for the inputs of the various feedback control loops that make up the automated system. A given programming command may specify the set point for the feedback loop, which in turn controls some action that the system is to accomplish. In effect, the purpose of the feedback loop is to verify that the programmed step has been carried out. For example, in a robot controller, the program might specify that the arm is to move to a designated position, and the feedback control system is used to verify that the move has been correctly made. The relationship of program control and feedback control in an automated system is illustrated in the figure below.

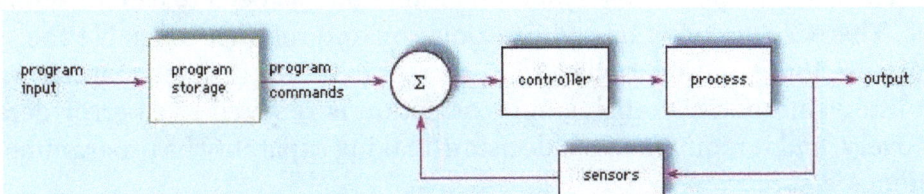

Relationship of program control and feedback control in an automated system

Some of the programmed commands may be executed in a simple open-loop fash-ion—i.e., without the need for a feedback loop to verify that the command has been properly carried out. For example, a command to flip an electrical switch may not require feedback. The need for feedback control in an automated system might arise when there are variations in the raw materials being fed into a production process, and the system must take these variations into consideration by making adjustments in its controlled actions. Without feedback, the system would be unable to exert sufficient control over the quality of the process output.

The programmed commands may be contained on mechanical devices (e.g., mechani-cal cams and linkages), punched paper tape, magnetic tape, magnetic disks, computer memory, or any of a variety of other media that have been developed over the years for particular applications. It is common today for automated equipment to use computer storage technology as the means for storing the programmed commands and convert-ing them into controlled actions. One of the advantages of computer storage is that the program can be readily changed or improved. Altering a program that is contained on mechanical cams involves considerable work.

Programmable machines are often capable of making decisions during their operation. The decision-making capacity is contained in the control program in the form of logical instructions that govern the operation of such a system under varying circumstances. Under one set of circumstances, the system responds one way; under different circum-stances, it responds in another way.

There are several reasons for providing an automated system with decision-making capability, including:

- Error detection and recovery,
- Safety monitoring,
- Interaction with humans, and
- Process optimization.

Error detection and recovery is concerned with decisions that must be made by the system in response to undesirable operating conditions. In the operation of any au-tomated system, malfunctions and errors sometimes occur during the normal cycle of operations, for which some form of corrective action must be taken to restore the system. The usual response to a system malfunction has been to call for human as-sistance. There is a growing trend in automation and robotics to enable the system itself to sense these malfunctions and to correct for them in some manner without human intervention. This sensing and correction is referred to as error detection and recovery, and it requires that a decision-making capability be programmed into the system.

Safety monitoring is a special case of error detection and recovery in which the malfunction involves a safety hazard. Decisions are required when the automated system sensors detect that a safety condition has developed that would be hazardous to the equipment or humans in the vicinity of the equipment. The purpose of the safety-monitoring system is to detect the hazard and to take the most appropriate action to remove or reduce it. This may involve stopping the operation and alerting maintenance personnel to the condition, or it may involve a more complex set of actions to eliminate the safety problem.

Automated systems are usually required to interact with humans in some way. An automatic bank teller machine, for example, must receive instructions from customers and act accordingly. In some automated systems, a variety of different instructions from humans is possible, and the decision-making capability of the system must be quite sophisticated in order to deal with the array of possibilities.

A fourth reason for decision making in an automated system is to optimize the process. The need for optimization occurs most commonly in processes in which there is an economic performance criterion whose optimization is desirable. For example, minimizing cost is usually an important objective in manufacturing. The automated system might use adaptive control to receive appropriate sensor signals and other inputs and make decisions to drive the process toward the optimal state.

## Open-loop and Closed-loop Control

Fundamentally, there are two types of control loop, open loop control, and closed loop feedback control.

In open loop control, the control action from the controller is independent of the "process output" (or "controlled process variable"). A good example of this is a central heating boiler controlled only by a timer, so that heat is applied for a constant time, regardless of the temperature of the building. (The control action is the switching on/off of the boiler. The process output is the building temperature).

In closed loop control, the control action from the controller is dependent on the process output. In the case of the boiler analogy this would include a thermostat to monitor the building temperature, and thereby feed back a signal to ensure the controller maintains the building at the temperature set on the thermostat. A closed loop controller therefore has a feedback loop which ensures the controller exerts a control action to give a process output the same as the "Reference input" or "set point". For this reason, closed loop controllers are also called feedback controllers.

The definition of a closed loop control system according to the British Standard Institution is 'a control system possessing monitoring feedback, the deviation signal formed as a result of this feedback being used to control the action of a final control element in such a way as to tend to reduce the deviation to zero.'

Likewise, a Feedback Control System is a system which tends to maintain a pre-scribed relationship of one system variable to another by comparing functions of these variables and using the difference as a means of control. The advanced type of automation that revolutionized manufacturing, aircraft, communications and oth-er industries, is feedback control, which is usually continuous and involves taking measurements using a sensor and making calculated adjustments to keep the mea-sured variable within a set range. The theoretical basis of closed loop automation is control theory.

A fly ball governor is an early example of a feedback control system.
An increase in speed would make the counterweights move outward,
sliding a linkage that tended to close the valve supplying steam, and so slowing the engine

## Control Actions

## Discrete Control

One of the simplest types of control is on-off control. An example is the thermostat used on household appliances which either opens or closes an electrical contact. (Thermo-stats were originally developed as true feedback-control mechanisms rather than the on-off common household appliance thermostat).

Sequence control, in which a programmed sequence of discrete operations is per-formed, often based on system logic that involves system states. An elevator control system is an example of sequence control.

## PID Controller

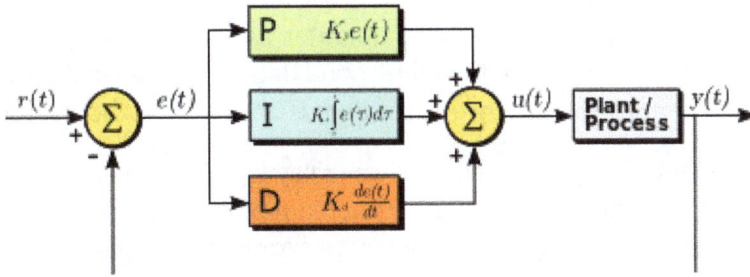

A block diagram of a PID controller in a feedback loop, r(t) is the desired process value or "set point", and y(t) is the measured process value.

A proportional–integral–derivative controller (PID controller) is a control loop feedback mechanism (controller) widely used in industrial control systems.

In a PID loop, the controller continuously calculates an error value as the difference between a desired set point and a measured process variable and applies a correction based on proportional, integral, and derivative terms, respectively (sometimes denoted P, I, and D) which give their name to the controller type.

The theoretical understanding and application dates from the 1920s, and they are implemented in nearly all analogue control systems; originally in mechanical controllers, and then using discrete electronics and latterly in industrial process computers.

## Sequential Control and Logical Sequence or System State Control

Sequential control may be either to a fixed sequence or to a logical one that will perform different actions depending on various system states. An example of an adjustable but otherwise fixed sequence is a timer on a lawn sprinkler.

## State Abstraction

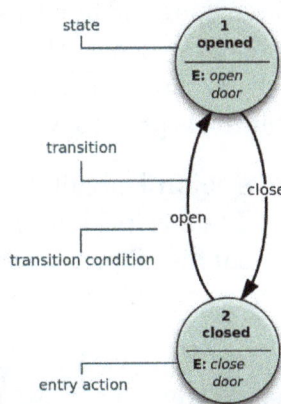

This state diagram shows how UML can be used for designing a
door system that can only be opened and closed

States refer to the various conditions that can occur in a use or sequence scenario of the system. An example is an elevator, which uses logic based on the system state to perform certain actions in response to its state and operator input. For example, if the operator presses the floor n button, the system will respond depending on whether the elevator is stopped or moving, going up or down, or if the door is open or closed, and other conditions.

An early development of sequential control was relay logic, by which electrical relays engage electrical contacts which either start or interrupt power to a device. Relays were first used in telegraph networks before being developed for controlling other devices, such as when starting and stopping industrial-sized electric motors or opening and closing solenoid valves. Using relays for control purposes allowed event-driven control, where actions could be triggered out of sequence, in response to external events. These were more flexible in their response than the rigid single-sequence cam timers. More complicated examples involved maintaining safe sequences for devices such as swing bridge controls, where a lock bolt needed to be disengaged before the bridge could be moved, and the lock bolt could not be released until the safety gates had already been closed.

The total number of relays, cam timers and drum sequencers can number into the hundreds or even thousands in some factories. Early programming techniques and languages were needed to make such systems manageable, one of the first being ladder logic, where diagrams of the interconnected relays resembled the rungs of a ladder. Special computers called programmable logic controllers were later designed to replace these collections of hardware with a single, more easily re-programmed unit.

In a typical hard wired motor start and stop circuit (called a control circuit) a motor is started by pushing a "Start" or "Run" button that activates a pair of electrical relays. The "lock-in" relay locks in contacts that keep the control circuit energized when the push button is released. (The start button is a normally open contact and the stop button is normally closed contact.) Another relay energizes a switch that powers the device that throws the motor starter switch (three sets of contacts for three phase industrial power) in the main power circuit. Large motors use high voltage and experience high in-rush current, making speed important in making and breaking contact. This can be dangerous for personnel and property with manual switches. The "lock in" contacts in the start circuit and the main power contacts for the motor are held engaged by their respective electromagnets until a "stop" or "off" button is pressed, which de-energizes the lock in relay.

Commonly interlocks are added to a control circuit. Suppose that the motor in the example is powering machinery that has a critical need for lubrication. In this case an interlock could be added to insure that the oil pump is running before the motor starts. Timers, limit switches and electric eyes are other common elements in control circuits.

Solenoid valves are widely used on compressed air or hydraulic fluid for powering actuators on mechanical components. While motors are used to supply continuous rotary motion, actuators are typically a better choice for intermittently creating a limited range of movement for a mechanical component, such as moving various mechanical arms, opening or closing valves, raising heavy press rolls, applying pressure to presses.

## Computer Control

Computers can perform both sequential control and feedback control, and typically a single computer will do both in an industrial application. Programmable logic controllers (PLCs) are a type of special purpose microprocessor that replaced many hardware components such as timers and drum sequencers used in relay logic type systems. General purpose process control computers have increasingly replaced stand alone controllers, with a single computer able to perform the operations of hundreds of controllers. Process control computers can process data from a network of PLCs, instruments and controllers in order to implement typical (such as PID) control of many individual variables or, in some cases, to implement complex control algorithms using multiple inputs and mathematical manipulations. They can also analyze data and create real time graphical displays for operators and run reports for operators, engineers and management.

Control of an automated teller machine (ATM) is an example of an interactive process in which a computer will perform a logic derived response to a user selection based on information retrieved from a networked database. The ATM process has similarities with other online transaction processes. The different logical responses are called scenarios. Such processes are typically designed with the aid of use cases and flowcharts, which guide the writing of the software code. The earliest feedback control mechanism was the water clock invented by Greek engineer Ctesibius.

## Advantages and Disadvantages

Perhaps the most cited advantage of automation in industry is that it is associated with faster production and cheaper labor costs. Another benefit could be that it replaces hard, physical, or monotonous work. Additionally, tasks that take place in hazardous environments or that are otherwise beyond human capabilities can be done by machines, as machines can operate even under extreme temperatures or in atmospheres that are radioactive or toxic. They can also be maintained with simple quality checks. However, at the time being, not all tasks can be automated, and some tasks are more expensive to automate than others. Initial costs of installing the machinery in factory settings are high, and failure to maintain a system could result in the loss of the product itself. Moreover, some studies seem to indicate that industrial automation could impose ill effects beyond operational concerns, including worker displacement due to systemic loss of employment and compounded environmental damage; however, these findings are both convoluted and controversial in nature, and could potentially be circumvented.

The main advantages of automation are:

- Increased throughput or productivity.

- Improved quality or increased predictability of quality.

- Improved robustness (consistency), of processes or product.

- Increased consistency of output.

- Reduced direct human labor costs and expenses.

- Installation in operations reduces cycle time.

- Can complete tasks where a high degree of accuracy is required.

- Replaces human operators in tasks that involve hard physical or monotonous work (e.g., using one forklift with a single driver instead of a team of multiple workers to lift a heavy object).

- Reduces some occupational injuries (e.g., fewer strained backs from lifting heavy objects).

- Replaces humans in tasks done in dangerous environments (i.e. fire, space, volcanoes, nuclear facilities, underwater, etc.).

- Performs tasks that are beyond human capabilities of size, weight, speed, endurance, etc.

- Reduces operation time and work handling time significantly.

- Frees up workers to take on other roles.

- Provides higher level jobs in the development, deployment, maintenance and running of the automated processes.

The main disadvantages of automation are:

- Possible security threats/vulnerability due to increased relative susceptibility for committing errors.

- Unpredictable or excessive development costs.

- High initial cost.

- Displaces workers due to job replacement.

- Leads to further environmental damage and could compound climate change.

## Societal Impact

Increased automation can often cause workers to feel anxious about losing their jobs as technology renders their skills or experience unnecessary. Early in the Industrial Revolution, when inventions like the steam engine were making some job categories expendable, workers forcefully resisted these changes. Luddites, for instance, were English textile workers who protested the introduction of weaving machines by destroying them. Similar movements have sprung up periodically ever since. For most of the nineteenth and twenti-eth centuries, the most influential of these movements were led by organized labor, which advocated for the retraining of workers whose jobs were rendered redundant by machines.

Currently, the relative anxiety about automation reflected in opinion polls seems to cor-relate closely with the strength of organized labor in that region or nation. For example, while a recent study by the Pew Research Center indicated that 72% of Americans are worried about increasing automation in the workplace, 80% of Swedes see automation and artificial intelligence as a good thing, due to the country's still-powerful unions and a more robust national safety net.

Automation is already contributing significantly to unemployment, particularly in na-tions where the government does not proactively seek to diminish its impact. In the United States, 47% of all current jobs have the potential to be fully automated by 2033, according to the research of experts Carl Benedikt Frey and Michael Osborne. Further-more, wages and educational attainment appear to be strongly negatively correlated with an occupation's risk of being automated. Prospects are particularly bleak for occu-pations that do not presently require a university degree, such as truck driving. Even in high-tech corridors like Silicon Valley, concern is spreading about a future in which a sizable percentage of adults have little chance of sustaining gainful employment. As the example of Sweden suggests, however, the transition to a more automated future need not inspire panic, if there is sufficient political will to promote the retraining of workers whose positions are being rendered obsolete.

## Lights Out Manufacturing

Lights out manufacturing is a production system with no human workers, to eliminate labor costs.

Lights out manufacturing grew in popularity in the U.S. when General Motors in 1982 implemented humans "hands-off" manufacturing in order to "replace risk-averse bu-reaucracy with automation and robots". However, the factory never reached full "lights out" status.

The expansion of Lights Out Manufacturing requires:

- Reliability of equipment;
- Long term mechanic capabilities;
- Planned preventative maintenance;
- Commitment from the staff.

## Health and Environment

The costs of automation to the environment are different depending on the technology, product or engine automated. There are automated engines that consume more energy resources from the Earth in comparison with previous engines and vice versa. Hazardous operations, such as oil refining, the manufacturing of industrial chemicals, and all forms of metal working, were always early contenders for automation.

The automation of vehicles could prove to have a substantial impact on the environment, although the nature of this impact could be beneficial or harmful depending on several factors. Because automated vehicles are much less likely to get into accidents compared to human-driven vehicles, some precautions built into current models (such as anti-lock brakes or laminated glass) would not be required for self-driving versions. Removing these safety features would also significantly reduce the weight of the vehicle, thus increasing fuel economy and reducing emissions per mile. Self-driving vehicles are also more precise with regard to acceleration and breaking, and this could contribute to reduced emissions. Self-driving cars could also potentially utilize fuel-efficient features such as route mapping that is able to calculate and take the most efficient routes. Despite this potential to reduce emissions, some researchers theorize that an increase of production of self-driving cars could lead to a boom of vehicle ownership and use. This boom could potentially negate any environmental benefits of self-driving cars if a large enough number of people begin driving personal vehicles more frequently.

Automation of homes and home appliances is also thought to impact the environment, but the benefits of these features are also questioned. A study of energy consumption of automated homes in Finland showed that smart homes could reduce energy consumption by monitoring levels of consumption in different areas of the home and adjusting consumption to reduce energy leaks (such as automatically reducing consumption during the nighttime when activity is low). This study, along with others, indicated that the smart home's ability to monitor and adjust consumption levels would reduce unnecessary energy usage. However, new research suggests that smart homes might not be as efficient as non-automated homes. A more recent study has indicated that, while monitoring and adjusting consumption levels does decrease unnecessary energy use, this process requires monitoring systems that also consume a significant amount of energy. This study suggested that the energy required to run these systems is so much so that it negates any benefits of the systems themselves, resulting in little to no ecological benefit.

## Convertibility and Turnaround Time

Another major shift in automation is the increased demand for flexibility and convertibility in manufacturing processes. Manufacturers are increasingly demanding the ability to easily switch from manufacturing Product A to manufacturing Product B without having to completely rebuild the production lines. Flexibility and distributed processes have led to the introduction of Automated Guided Vehicles with Natural Features Navigation.

Digital electronics helped too. Former analogue-based instrumentation was replaced by digital equivalents which can be more accurate and flexible, and offer greater scope for more sophisticated configuration, parameterization and operation. This was accompanied by the fieldbus revolution which provided a networked (i.e. a single cable) means of communicating between control systems and field level instrumentation, eliminating hard-wiring.

Discrete manufacturing plants adopted these technologies fast. The more conservative process industries with their longer plant life cycles have been slower to adopt and analogue-based measurement and control still dominates. The growing use of Industrial Ethernet on the factory floor is pushing these trends still further, enabling manufacturing plants to be integrated more tightly within the enterprise, via the internet if necessary. Global competition has also increased demand for Reconfigurable Manufacturing Systems.

## Automation Tools

Engineers can now have numerical control over automated devices. The result has been a rapidly expanding range of applications and human activities. Computer-aided technologies (or CAx) now serve as the basis for mathematical and organizational tools used to create complex systems. Notable examples of CAx include Computer-aided design (CAD software) and Computer-aided manufacturing (CAM software). The improved design, analysis, and manufacture of products enabled by CAx has been beneficial for industry.

Information technology, together with industrial machinery and processes, can assist in the design, implementation, and monitoring of control systems. One example of an industrial control system is a programmable logic controller (PLC). PLCs are specialized hardened computers which are frequently used to synchronize the flow of inputs from (physical) sensors and events with the flow of outputs to actuators and events.

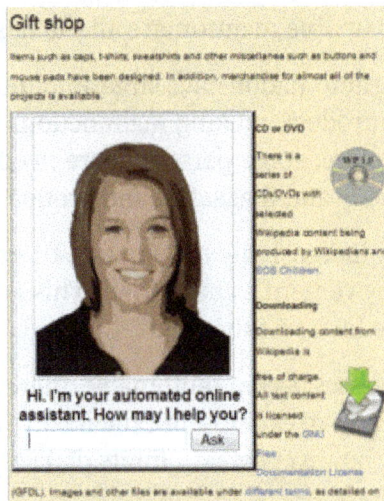

An automated online assistant on a website, with an avatar for enhanced human–computer interaction.

Human-machine interfaces (HMI) or computer human interfaces (CHI), formerly known as man-machine interfaces, are usually employed to communicate with PLCs and other computers. Service personnel who monitor and control through HMIs can be called by different names. In industrial process and manufacturing environments, they are called operators or something similar. In boiler houses and central utilities departments they are called stationary engineers.

Different types of automation tools exist:

- ANN – Artificial neural network
- DCS – Distributed Control System
- HMI – Human Machine Interface
- SCADA – Supervisory Control and Data Acquisition
- PLC – Programmable Logic Controller
- Instrumentation
- Motion control
- Robotics.

When it comes to factory automation, Host Simulation Software (HSS) is a commonly used testing tool that is used to test the equipment software. HSS is used to test equipment performance with respect to Factory Automation standards (timeouts, response time, processing time).

## Limitations to Automation

- Current technology is unable to automate all the desired tasks.
- Many operations using automation have large amounts of invested capital and produce high volumes of product, making malfunctions extremely costly and potentially hazardous. Therefore, some personnel are needed to ensure that the entire system functions properly and that safety and product quality are maintained.
- As a process becomes increasingly automated, there is less and less labor to be saved or quality. improvement to be gained. This is an example of both diminishing returns and the logistic function.
- As more and more processes become automated, there are fewer remaining non-automated processes. This is an example of exhaustion of opportunities. New technological paradigms may however set new limits that surpass the previous limits.

## Current Limitations

Many roles for humans in industrial processes presently lie beyond the scope of automation. Human-level pattern recognition, language comprehension, and language production ability are well beyond the capabilities of modern mechanical and computer systems. Tasks requiring subjective assessment or synthesis of complex sensory data, such as scents and sounds, as well as high-level tasks such as strategic planning, currently require human expertise. In many cases, the use of humans is more cost-effective than mechanical approaches even where automation of industrial tasks is possible. Overcoming these obstacles is a theorized path to post-scarcity economics.

## Paradox of Automation

The paradox of automation says that the more efficient the automated system, the more crucial the human contribution of the operators. Humans are less involved, but their involvement becomes more critical.

If an automated system has an error, it will multiply that error until it's fixed or shut down. This is where human operators come in.

A fatal example of this was Air France Flight 447, where a failure of automation put the pilots into a manual situation they were not prepared for.

## Cognitive Automation

Cognitive automation is an emerging genus of automation enabled by cognitive computing. Its primary concern is the automation of clerical tasks and workflows that consist of structuring unstructured data.

Cognitive automation relies on multiple disciplines: natural language processing, real-time computing, machine learning algorithms, big data analytics and evidence-based learning. According to Deloitte, cognitive automation enables the replication of human tasks and judgment "at rapid speeds and considerable scale."

Such tasks include:

- Document redaction

- Data extraction and document synthesis / reporting

- Contract management

- Natural language search

- Customer, employee, and stakeholder onboarding

- Manual activities and verifications

- Follow up and email communications.

## Recent and Emerging Applications

KUKA industrial robots being used at a bakery for food production

## Automated Retail

## Food and Drink

The food retail industry has started to apply automation to the ordering process; McDonald's has introduced touch screen ordering and payment systems in many of its restaurants, reducing the need for as many cashier employees. The University of Texas at Austin has introduced fully automated cafe retail locations. Some Cafes and restaurants have utilized mobile and tablet "apps" to make the ordering process more efficient by customers ordering and paying on their device. Some restaurants have automated food delivery to customers tables using a Conveyor belt system. The use of robots is sometimes employed to replace waiting staff.

## Stores

Many supermarkets and even smaller stores are rapidly introducing Self checkout systems reducing the need for employing checkout workers. In the United States, the retail industry employs 15.9 million people as of 2017 (around 1 in 9 Americans in the workforce). Globally, an estimated 192 million workers could be affected by automation according to research by Eurasia Group.

Online shopping could be considered a form of automated retail as the payment and checkout are through an automated Online transaction processing system, with the share of online retail accounting jumping from 5.1% in 2011 to 8.3% in 2016.

However, two-thirds of books, music and films are now purchased online. In addition, automation and online shopping could reduce demands for shopping malls, and retail property, which in America is currently estimated to account for 31% of all commercial property or around 7 billion square feet. Amazon has gained much of the growth in recent years for online shopping, accounting for half of the growth in online retail in 2016. Other forms of automation can also be an integral part of online shopping, for example the deployment of automated warehouse robotics such as that applied by Amazon using Kiva Systems.

## Automated Mining

Automated mining involves the removal of human labor from the mining process. The mining industry is currently in the transition towards automation. Currently it can still require a large amount of human capital, particularly in the third world where labor costs are low so there is less incentive for increasing efficiency through automation.

## Automated Video Surveillance

The Defense Advanced Research Projects Agency (DARPA) started the research and development of automated visual surveillance and monitoring (VSAM) program, between 1997 and 1999, and airborne video surveillance (AVS) programs, from 1998 to 2002. Currently, there is a major effort underway in the vision community to develop a fully automated tracking surveillance system. Automated video surveillance monitors people and vehicles in real time within a busy environment. Existing automated surveillance systems are based on the environment they are primarily designed to observe, i.e., indoor, outdoor or airborne, the amount of sensors that the automated system can handle and the mobility of sensor, i.e., stationary camera vs. mobile camera. The purpose of a surveillance system is to record properties and trajectories of objects in a given area, generate warnings or notify designated authority in case of occurrence of particular events.

## Automated Highway Systems

As demands for safety and mobility have grown and technological possibilities have multiplied, interest in automation has grown. Seeking to accelerate the development and introduction of fully automated vehicles and highways, the United States Congress authorized more than $650 million over six years for intelligent transport systems (ITS) and demonstration projects in the 1991 Intermodal Surface Transportation Efficiency Act (ISTEA). Congress legislated in ISTEA that "the Secretary of Transportation shall develop an automated highway and vehicle prototype from which future fully automated intelligent vehicle-highway systems can be developed. Such development shall include research in human factors to ensure the success of the man-machine relationship. The goal of this program is to have the first fully automated highway roadway or an automated test track in operation by 1997. This system shall accommodate installation of equipment in new and existing motor vehicles."

Full automation commonly defined as requiring no control or very limited control by the driver; such automation would be accomplished through a combination of sensor, computer, and communications systems in vehicles and along the roadway. Fully automated driving would, in theory, allow closer vehicle spacing and higher speeds, which could enhance traffic capacity in places where additional road building is physically impossible, politically unacceptable, or prohibitively expensive. Automated controls also might enhance road safety by reducing the opportunity for driver error, which causes a large share of motor vehicle crashes. Other potential benefits include improved air quality (as a result of more-efficient traffic flows), increased fuel economy, and spin-off technologies generated during research and development related to automated highway systems.

## Automated Waste Management

Automated side loader operation

Automated waste collection trucks prevent the need for as many workers as well as easing the level of labor required to provide the service.

## Home Automation

Home automation (also called domotics) designates an emerging practice of increased automation of household appliances and features in residential dwellings, particularly through electronic means that allow for things impracticable, overly expensive or simply not possible in recent past decades.

## Laboratory Automation

Automation is essential for many scientific and clinical applications. Therefore, automation has been extensively employed in laboratories. From as early as 1980 fully automated laboratories have already been working. However, automation has not become widespread in laboratories due to its high cost. This may change with the ability of integrating low-cost devices with standard laboratory equipment. Auto samplers are common devices used in laboratory automation.

## Industrial Automation

Industrial automation deals primarily with the automation of manufacturing, quality control and material handling processes. General purpose controllers for industrial processes include Programmable logic controllers, stand-alone I/O modules, and computers. Industrial automation is to replace the decision making of humans and manual command-response activities with the use of mechanized equipment and logical programming commands. One trend is increased use of Machine vision to provide automatic inspection and robot guidance functions, another is a continuing increase in the use of robots. Industrial automation is simply require in industries.

The integration of control and information across the enterprise enables industries to optimise industrial process operations.

Energy efficiency in industrial processes has become a higher priority. Semiconductor companies like Infineon Technologies are offering 8-bit micro-controller applications for example found in motor controls, general purpose pumps, fans, and e-bikes to reduce energy consumption and thus increase efficiency.

## Industrial Automation and Industry

The rise of industrial automation is directly tied to the "fourth industrial revolution", which is better known now as Industry 4.0. Originating from Germany, Industry 4.0 encompasses numerous devises, concepts, and machines . It, along with the advancement of the Industrial Internet of Things (formally known as the IoT or IIoT) which is "Internet of Things is a seamless integration of diverse physical objects in the Internet through a virtual representation". These new revolutionary advancements have drawn attention to the world of automation in an entirely new light and shown ways for it to grow to increase productivity and efficiency in machinery and manufacturing facilities. Industry 4.0 works with the IIoT and software/hardware to connect in a way that (through communication technologies) add enhancements and improve manufacturing processes. Being able to create smarter, safer, and more advanced manufacturing is now possible with these new technologies. It opens up a manufacturing platform that is more reliable, consistent, and efficient that before. Implementation of systems such as SCADA are an example of software that take place in Industrial Automation today.

SCADA is a supervisory data collection software, just one of the many used in Industrial Automation . Industry 4.0 vastly covers many areas in manufacturing and will continue to do so as time goes on.

## Industrial Robotics

Industrial robotics is a sub-branch in the industrial automation that aids in various manufacturing processes. Such manufacturing processes include; machining, welding, painting, assembling and material handling to name a few. Industrial robots utilizes various mechanical, electrical as well as software systems to allow for high precision, accuracy and speed that far exceeds any human performance. The birth of industrial robot came shortly after World War II as United States saw the need for a quicker way to produce industrial and consumer goods. Servos, digital logic and solid state electronics allowed engineers to build better and faster systems and overtime these systems were improved and revised to the point where a single robot is capable of running 24 hours a day with little or no maintenance. In 1997, there were 700,000 industrial robots in use, the number has risen to 1.8M in 2017.

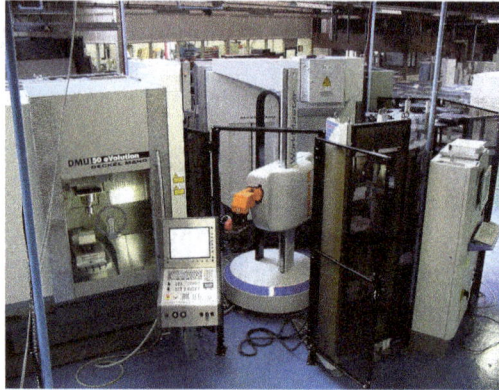

Automated milling machines

## Programmable Logic Controllers

Siemens Simatic S7-400 system in a rack, left-to-right: power supply unit (PSU), CPU, interface module (IM) and communication processor (CP)

Industrial automation incorporates programmable logic controllers in the manufacturing process. Programmable logic controllers (PLCs) use a processing system which allows for variation of controls of inputs and outputs using simple programming. PLCs make use of programmable memory, storing instructions and functions like logic, sequencing, timing, counting, etc. Using a logic based language, a PLC can receive a variety of inputs and return a variety of logical outputs, the input devices being sensors and output devices being motors, valves, etc. PLCs are similar to computers, however, while computers are optimized for calculations, PLCs are optimized for control task and use in industrial environments. They are built so that only basic logic-based programming knowledge is needed and to handle vibrations, high temperatures, humidity and noise. The greatest advantage PLCs offer is their flexibility. With the same basic controllers, a PLC can operate a range of different control systems. PLCs make it unnecessary to rewire a system to change the control system. This flexibility leads to a cost-effective system for complex and varied control systems.

PLCs can range from small "building brick" devices with tens of I/O in a housing integral with the processor, to large rack-mounted modular devices with a count of thousands of I/O, and which are often networked to other PLC and SCADA systems.

They can be designed for multiple arrangements of digital and analog inputs and outputs (I/O), extended temperature ranges, immunity to electrical noise, and resistance to vibration and impact. Programs to control machine operation are typically stored in battery-backed-up or non-volatile memory.

It was from the automotive industry in the USA that the PLC was born. Before the PLC, control, sequencing, and safety interlock logic for manufacturing automobiles was mainly composed of relays, cam timers, drum sequencers, and dedicated closed-loop controllers. Since these could number in the hundreds or even thousands, the process for updating such facilities for the yearly model change-over was very time consuming and expensive, as electricians needed to individually rewire the relays to change their operational characteristics.

When digital computers became available, being general-purpose programmable devices, they were soon applied to control sequential and combinatorial logic in industrial processes. However these early computers required specialist programmers and stringent operating environmental control for temperature, cleanliness, and power quality. To meet these challenges this the PLC was developed with several key attributes. It would tolerate the shop-floor environment, it would support discrete (bit-form) input and output in an easily extensible manner, it would not require years of training to use, and it would permit its operation to be monitored. Since many industrial processes have timescales easily addressed by millisecond response times, modern (fast, small, reliable) electronics greatly facilitate building reliable controllers, and performance could be traded off for reliability.

## Agent-assisted Automation

Agent-assisted automation refers to automation used by call center agents to handle customer inquiries. There are two basic types: desktop automation and automated voice solutions. Desktop automation refers to software programming that makes it easier for the call center agent to work across multiple desktop tools. The automation would take the information entered into one tool and populate it across the others so it did not have to be entered more than once, for example. Automated voice solutions allow the agents to remain on the line while disclosures and other important information is provided to customers in the form of pre-recorded audio files. Specialized applications of these automated voice solutions enable the agents to process credit cards without ever seeing or hearing the credit card numbers or CVV codes.

The key benefit of agent-assisted automation is compliance and error-proofing. Agents are sometimes not fully trained or they forget or ignore key steps in the process. The use of automation ensures that what is supposed to happen on the call actually does, every time.

# Industrial Automation

To resolve the automation and control issues, industries use the ever changing technologies in control systems for efficient production or manufacturing processes. These requires the high quality and reliable control systems. New trends in industrial automation deals with latest control devices and communication protocols to control field devices like control valves and other final control elements. Some of the smart devices or instruments used in automated industry has the ability to control the processes and also communication capabilities without interfacing to other field level control devices like PLC's.

Industrial automation is the use of various control devices like PC's/PLC's/DCS, used to have control on various operations of an industry without significant intervention from humans and to provide automatic control performance. In industries, control strategies use a set of technologies which are implemented to get the desired performance or output, making the automation system most essential for industries.

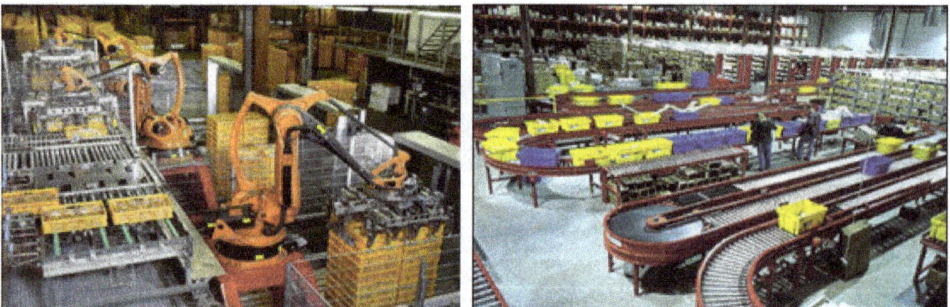

Automated process in industries

Industrial automation involves usage of advanced control strategies like cascade controls, modern control hardware devices as PLC's, sensors and other instruments for sensing the control variables, signal conditioning equipments to connect the signals to the control devices, drives and other significant final control devices, standalone computing systems, communication systems, alarming and HMI (Human Machine Interface) systems.

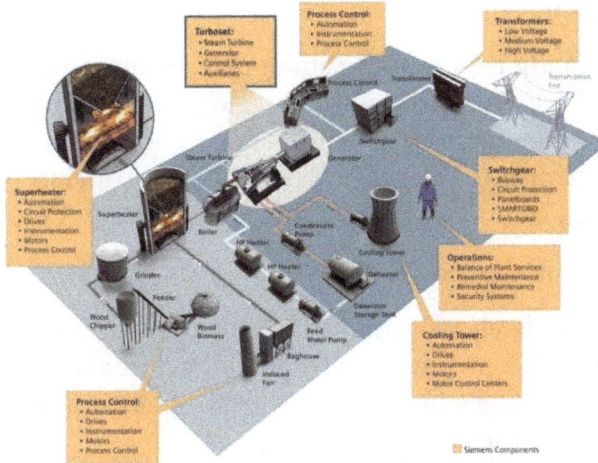

The above figure shows the power plant automation provided by Siemens for achieving sustainable, safe and economic operations. It provides the total integrated automation (TIA) by automating every section of power plant with efficient control devices, field sensors and actuating devices. In this automation, SIMATIC modules (PLCs) are used as control devices while Win-CC provides an effective graphical interface.

## Advantages of Automation System

- To increase productivity

  Automation of factory or manufacturing or process plant improves production rate through a better control of production. It helps to produce mass production by drastically reducing assembly time per product with a greater production quality. Therefore, for a given labor input it produces a large amount of output.

- To provide optimum cost of operation

  Integration of various processes in industry with automated machineries, minimizes cycle times and effort and hence the need of human labor gets reduced. Thus the investment on employees has been saved with automation.

- To improve product quality

  Since the automation reduces the human involvement, the possibility of human errors also gets eliminated. Uniformity and product quality with a greater conformity can be maintained with automation by adaptively controlling and monitoring the industrial processes in all stages right from inception of a product to an end product.

- To reduce routine checks

  Automation completely reduces the need for manual checking of various process parameters. By taking advantage of automation technologies, industrial processes automatically adjusts process variables to set or desired values using closed loop control techniques.

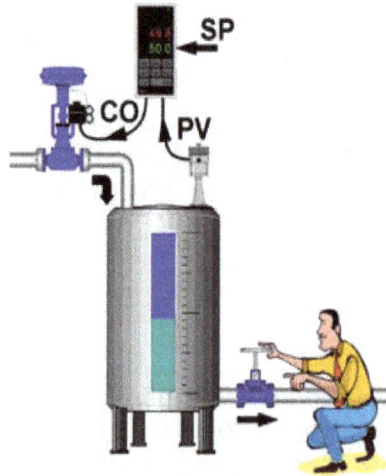

- To raise the level of safety

  Industrial automation increases the level of safety to personnel by substituting them with automated machines in hazardous working conditions. Traditionally, industrial robots and robotic devices are implemented in such risky and hazardous places.

## Control Loop

In today's modern plants, processes are controlled to achieve desired objectives. The term control means methods or means used to force parameters in the environment to have specific or desired values. To achieve control, several elements are coordinated together to achieve the control objective. All the elements necessary to accomplish the control objectives, including the instrumentation systems, are usually described by the term control system.

Control can either be manual or automatic. Manual control involves human intervention and it often entails forcing a given parameter to specific value with the human factor actually carrying out the control function. For example, suppose we want to control the level of water in an open tank which has an inlet valve through which water passes, all we simply do is to station an operator who basically uses his judgment to close the valve when the tank has become full or open the valve when the tank is almost empty.

In automatic control, no human intervention is required rather sensors, controllers, actuators and other control elements are used to automatically control a system to force the system parameters to desired levels.

## Elements of an Automatic Control Loop

An automatic control loop is made up of the following elements:

## Process

In general, a process can consist of a complex assembly of phenomena that relates to some manufacturing sequence or any system we wish to control. Many variables may be involved in such a process, and it may be desirable to control all these variables at the same time. There are single-variable processes, in which only one variable is to be controlled, as well as multi-variable processes, in which many variables, perhaps interrelated, may require regulation.

## Measurement

To achieve the control of a variable in a process, we must have information on the variable itself. Such information is found by measuring the variable. In general, a measurement refers to the conversion of the variable into some corresponding analog signal of the variable, such as a pneumatic pressure, an electrical voltage, or current. The result of the measurement is a conversion of the variable into some proportional information in a useful form required by the other elements in the process control operation.

## Sensors/Transducers

A sensor is a device that performs the initial measurement and energy conversion of a variable into analogous electrical or pneumatic information. Sometimes further transformation or signal conditioning may be required to complete the measurement function. The sensor used for measurement may also be called a transducer The word sensor is preferred for the initial measurement device, while a "transducer" represents a device that converts any signal from one form to another. Thus, for example, a device that converts a voltage into a proportional current would be a transducer. In other words, all sensors are transducers, but not all transducers are sensors.

## Error Detector

This is the device that determines whether the variable we desire to control, often called the process variable is above or below the desired level called the Set point or reference value. If the process variable is above or below the set point, an error signal proportional to the error is generated. This error signal is then used by the controller to generate a control action. So before any control action takes place, an error signal must be generated. It is worthy to note that the error detector is often an integral part of the controller device; however it is important to keep a clear distinction between the two.

## Controller

The device that acts on the error signal generated to determine what control action, if any, to be taken is called a controller. The evaluation performed to determine control action can be done by electronic signal processing, by pneumatic signal processing, or by a computer. Computer use is growing rapidly in the field of process control because computers are easily adapted to the decision-making operations and because of their inherent capacity to handle control of multivariable systems. The controller requires an input of both a measured indication of the controlled variable and a representation of the reference value of the variable, expressed in the same terms as the measured value. The reference value of the variable, you will recall, is referred to as the set point. Evaluation consists of determining action required to bring the controlled variable to the set point value.

## Control Element

The final element in the control loop is a control element that exerts a direct influence on the process; it is the device that provides those required changes in the controlled variable to bring it to the setpoint. This element accepts an input from the controller, which is then transformed into some proportional operation performed on the process. In most process control loops, the final control element is a valve which is often referred to as the final control element.

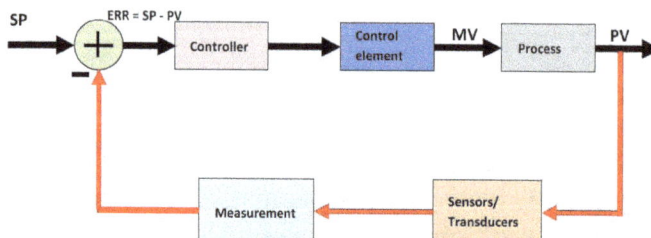

Shown above is the block diagram of a typical process control loop with feedback control

In a control loop, the signal flow forms a complete circuit from the process through measurement, error detector, controller, and final control element. This is called a loop, and in general we speak of a process control loop. In most cases this is called a feedback loop, because we determine an error and feed back a correction to the process. We also have open loop. In open loop , there is no feedback from the process. Here, the control action does not depend on changes in the process variable. Control action is usually based on changes in the input to the process.

## Working of a Process Control Loop

It is important to understand how a process control loop works, what control is and what a control system does. Having now understood these basic concepts, how then does a process control loop works? To understand how it works, we refer to the above block diagram of a feedback control loop.

The first point of interest for any process control endeavor is the process variable, PV. It is the variable we have chosen to control or maintain at a given reference value or set point. So as shown in the process control loop above, the process variable, PV, is measured with the aid of sensors/transducers. This measured signal is then fed into a controller incorporating an error detector device. Here, the process variable, PV is compared with the desired value of the process variable or the set point, SP and an error signal with a specific magnitude and polarity is generated and further processed within the controller. Based on the processed error signal, the controller initiates a control action with the aid of the control element or final control element as it is often known. The final control element, initiates a change in the process by changing the manipulated variable, MV.

Which then alters the process until it settles at the set point. In this way, the process variable is taken back to its desired value or set point. This is essentially how a process control loop works. Most complex process plants are operated with this simple underlying principle of process control.

## Programmable Logic Controller

A Programmable Logic Controller, or PLC, is a ruggedized computer used for industrial automation. These controllers can automate a specific process, machine function, or even an entire production line.

The image above is an Allen-Bradley PLC rack, a common example of a PLC setup that includes a CPU, analog inputs, analog outputs, and dc outputs.

A programmable logic controller (PLC) is an industrial solid-state computer that monitors inputs and outputs, and makes logic-based decisions for automated processes or machines.

PLCs were introduced in the late 1960s by inventor Richard Morley to provide the same functions as relay logic systems. Relay systems at the time tended to fail and create delays. Technicians then had to troubleshoot an entire wall of relays to fix the problem.

PLCs are robust and can survive harsh conditions including severe heat, cold, dust, and extreme moisture. Their programming language is easily understood, so they can be programmed without much difficulty. PLCs are modular so they can be plugged into various setups. Relays switching under load can cause undesired arcing between contacts. Arcing generates high temperatures that weld contacts shut and cause degradation of the contacts in the relays, resulting in device failure. Replacing relays with PLCs helps prevent overheating of contacts.

PLCs do have disadvantages. They do not perform well when handling complex data. When dealing with data that requires C++ or Visual Basic, computers are the controllers of choice. PLCs also cannot display data well, so external monitors are often required.

## PLC Hardware Components

### PLC System

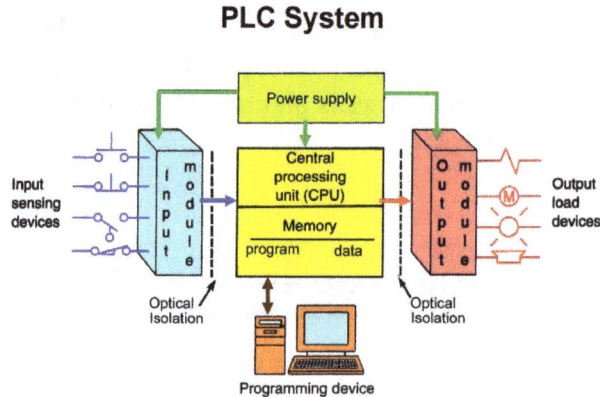

PLCs work with inputs, outputs, a power supply, and external programming devices

A central processing unit (CPU) serves as the brain of the PLC. It is a -16 or -32 bit microprocessor consisting of a memory chip and integrated circuits for control logic, monitoring, and communicating. The CPU directs the PLC to execute control instructions, communicate with other devices, carry out logic and arithmetic operations, and perform internal diagnostics. The CPU runs memory routines, constantly checking the PLC (PLC controller is redundant) to avoid programming errors and ensure the memory is undamaged.

Memory provides permanent storage to the operating system for data used by the CPU. The system's read-only memory (ROM) stores data permanently for the operating system random access memory (RAM) stores status information for input and output devices, along with values for timers, counters, and internal devices. PLCs require a programming device, either a computer or console, to upload data onto the CPU.

## CPU Operating Cycle

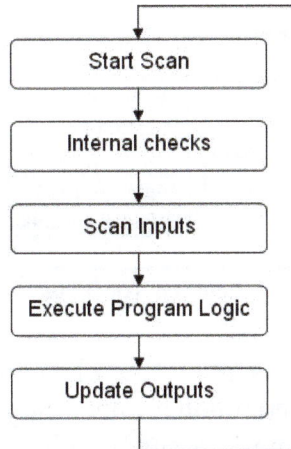

A CPU operating cycle includes the following steps: a) start scan; b) internal checks; c) scan inputs; d) execute program logic; and e) update outputs. The program repeats with the updated outputs

PLCs read signals from different sensors and input devices. These input devices can be keyboards, switches, or sensors. Inputs can be either in digital or analog form. Robots and visual systems are intelligent devices that can send signals to PLC input modules. Output devices such as motors and solenoid valves complete the automated system.

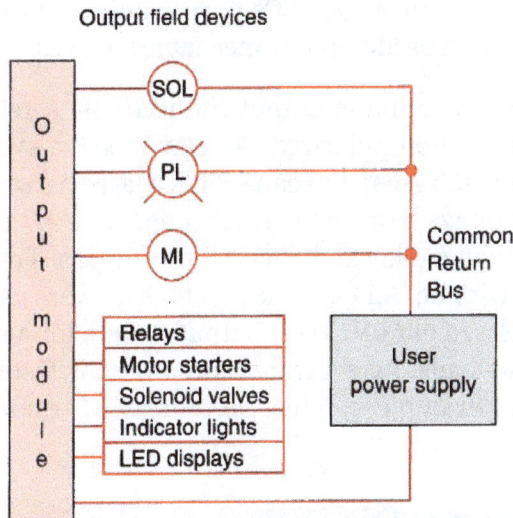

The top image depicts common inputs in a PLC, including push buttons and switches. Output connections are shown in the bottom image and include signal out (SOL), pilot light (PL), and motor ignition (MI)

Input field devices

Common
Power
Bus

| Relay contacts |
| User power supply | Pushbuttons |
| | Limit switches |
| | Analog sensors |
| | Selector switches |

Input module

Common Return Bus

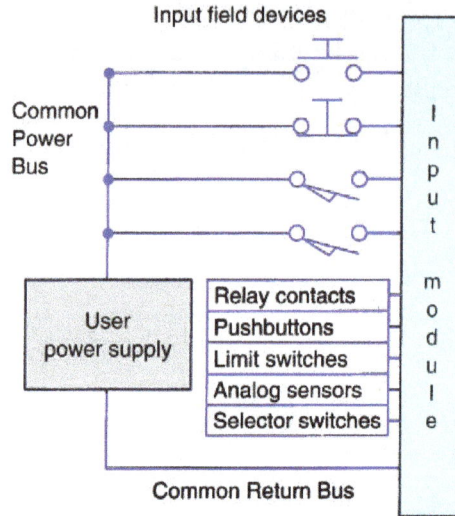

Sinking and sourcing are two important terms when discussing input and output connections of PLCs. Sinking is the common ground line (-) and sourcing is the common VCC line (+). VCC stands for the positive supply voltage connection point. Sinking and sourcing inputs only conduct electricity in one direction. Each input has its own return line, and several inputs connect to one return line instead of several separate return lines. These common lines are labeled "COMM." Sensor outputs mark the size of the signal given.

Direct current (dc) input modules connect to sourcing or sinking transistor type devices. Alternating current (ac) input modules are less common than dc inputs because most sensors have transistor outputs, so if the system uses a sensor input, it will most likely be dc; ac inputs take longer for PLCs to see compared to dc inputs. A typical ac input is a mechanical switch used for slow mechanical drives.

Relays are one of the most common output connections. A relay can switch ac or dc modules because they are non-polarized. A relay is slow, switching and settling at speeds of 5 to 50 milliseconds (ms), but can switch a large current. For example, a relay can be used for a low-voltage battery to switch a 230 volt AC main circuit. Transistor connections are faster than a relay and have a long lifespan. Transistors switch a small current, but only work with dc. An example of a high-power transistor has a current of 15 amps with a max voltage of 60V. Triac output (triode for alternating current) connections only control ac loads. Like a transistor, a triac is faster and handles large ac loads. A triac output, for example, can handle voltages of 500 to 800 with a current of 12 amps.

## Working of a Programmable Logic Controller

There are four basic steps in the operation of all PLCs; Input Scan, Program Scan, Output Scan, and Housekeeping. These steps continually take place in a repeating loop.

Four steps in the PLC operations:

1.  Input Scan

    - Detects the state of all input devices that are connected to the PLC

2.  Program Scan

    - Executes the user created program logic

3.  Output Scan

    - Energizes or de-energize all output devices that are connected to the PLC.

4.  Housekeeping

    - This step includes communications with programming terminals, internal diagnostics, etc.

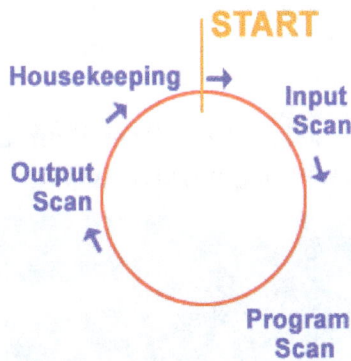

These steps are continually processed in a loop

## Types of PLCs

Programmable Logic Controllers (PLCs) are integrated as either single or modular units.

An integrated or Compact PLC is built by several modules within a single case. Therefore, the I/O capabilities are decided by the manufacturer, but not by the user. Some of the integrated PLCs allow to connect additional I/O to make them somewhat modular.

## Integrated or Compact PLCs

A modular PLC is built with several components that are plugged into a common rack or bus with extendable I/O capabilities. It contains power supply module, CPU and other I/O modules that are plugged together in the same rack, which are from same manufacturers or from other manufacturers. These modular PLCs come in different sizes with variable power supply, computing capabilities, I/O connectivity, etc.

A modular Types of PLC

Modular PLCs are further divided into small, medium and large PLCs based on the program memory size and the number of I/O features.

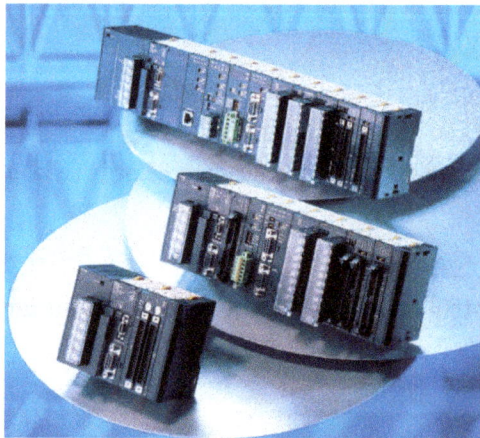

Small, Medium and Large sized Types of PLCs

Small PLC is a mini-sized PLC that is designed as compact and robust unit mounted or placed beside the equipment to be controlled. This type of PLC is used for replacing hard-wired relay logics, counters, timers, etc. This PLC I/O module expandability is limited for one or two modules and it uses logic instruction list or relay ladder language as programming language.

Medium-sized PLC is mostly used PLC in industries which allows many plug-in modules that are mounted on backplane of the system. Some hundreds of input/ output points are provided by adding additional I/O cards – and, in addition to these – communication module facilities are provided by this PLC.

Large PLCs are used wherein complex process control functions are required. These PLCs' capacities are quite higher than the medium PLCs in terms of memory, programming languages, I/O points, and communication modules, and so on. Mostly, these PLCs are used in supervisory control and data acquisition (SCADA) systems, larger plants, distributed control systems, etc.

## Input / Output Devices

– Switches and Pushbuttons
– Sensing Devices

• Limit Switches
• Photoelectric Sensors
• Proximity Sensors

– Valves
– Motor Starters
– Solenoids
– Actuators

– Condition Sensors
– Encoders

• Pressure Switches
• Level Switches
• Temperature Switches
• Vacuum Switches
• Float Switches

– Horns and Alarms
– Stack lights
– Control Relays
– Counter/Totalizer
– Pumps – Printers
– Fans

## Things to Consider when Choosing a PLC

There are many PLC systems on the market today. Other than cost, you must consider the following when deciding which one will best suit the needs of your application:

• Will the system be powered by AC or DC voltage

• Does the PLC have enough memory to run my user program

• Does the system run fast enough to meet my application's requirements

• What type of software is used to program the PLC

• Will the PLC be able to manage the number of inputs and outputs that my application requires

• If required by your application, can the PLC handle analog inputs and outputs, or maybe a combination of both analog and discrete inputs and outputs

• How am I going to communicate with my PLC

- Do I need network connectivity and can it be added to my PLC

- Will the system be located in one place or spread out over a large area.

## PLC Applications

The simple suitable application is a conveyor system. The requirements of the conveyor systems are as follows:

- A programmable logic controller is used to start and stop the motors of the conveyor belt.

- The conveyor system has three segmented conveyor belts. Each segment is run by a motor.

- To detect the position of a plate, a proximity switch is positioned at the segment's end.

- The first conveyor segment is turned ON always.

- The proximity switch in the first segment detects the plate to turn ON the second conveyor segment.

- The third conveyor segment is turned ON when the proximity switch detects the plate at the second conveyor.

- As the plate comes out of the detection range, the second conveyor is stopped after 20 sec.

- When the proximity switch fails to detect the plate, the third conveyor is stopped after 20 sec.

## Advantages

- PLCs can be programmed easily which can be understood clearly well.

- They are fabricated to survive vibrations, noise, humidity, and temperature.

- The controller has the input and output for interfacing.

## Disadvantages

- It is a tedious job when replacing or bringing any changes to it.

- Skillful work force is required to find its errors.

- Lot of effort is put to connect the wires.

- The hold up time is usually indefinite when any problem arises.

# Human Machine Interface

A Human-Machine Interface (HMI) is a user interface or dashboard that connects a person to a machine, system, or device. While the term can technically be applied to any screen that allows a user to interact with a device, HMI is most commonly used in the context of an industrial process.

Although HMI is the most common term for this technology, it is sometimes referred to as Man-Machine Interface (MMI), Operator Interface Terminal (OIT), Local Operator Interface (LOI), or Operator Terminal (OT). HMI and Graphical User Interface (GUI) are similar but not synonymous: GUIs are often leveraged within HMIs for visualization capabilities.

In industrial settings, HMIs can be used to:

- Visually display data
- Track production time, trends, and tags
- Oversee KPIs
- Monitor machine inputs and outputs.

Similar to how you would interact with your air-conditioning system to check and control the temperature in your house, a plant-floor operator might use an HMI to check and control the temperature of an industrial water tank, or to see if a certain pump in the facility is currently running.

HMIs come in a variety of forms, from built-in screens on machines, to computer monitors, to tablets, but regardless of their format or which term you use to refer to them, their purpose is to provide insight into mechanical performance and progress.

## Basic Types of HMIs

There are three basic types of HMIs: the push button replacer, the data handler, and the overseer. Before the HMI came into existence, a control might consist of hundreds of pushbuttons and LEDs performing different operations. The pushbutton replacer HMI has streamlined manufacturing processes, centralizing all the functions of each button into one location. The data handler is perfect for applications requiring constant feedback from the system, or printouts of the production reports. With the data handler, you must ensure the HMI screen is big enough for such things as graphs, visual representations and production summaries. The data handler includes such functions as recipes, data trending, data logging and alarm handling/logging. Finally, anytime an application involves SCADA or MES, an overseer HMI is extremely beneficial. The overseer HMI will most likely need to run Windows, and have several Ethernet ports.

## Selecting an HMI

An HMI is a substantial purchase, so it is important to know exactly what is required of it. An HMI is used for three primary roles: a pushbutton replacer, data handler, and overseer. The pushbutton replacer takes the place of LEDs, On/ Off buttons, switches or any mechanical device that performs a control function. The elimination of these mechanical devices is possible because the HMI can provide a visual representation of all these devices on its LCD screen, while performing all the same functions. The Data Handler is used for applications that require constant feedback and monitoring. Often these Data Handlers come equipped with large capacity memories. The last of the HMI three types is referred to as the overseer, because it works with SCADA and MES. These are centralized systems that monitor and control entire sites or complexes of large systems spread out over large areas. An HMI is usually linked to the SCADA system's databases and software programs, to provide trending, diagnostic data, and management information.

## Physical Properties of a HMI

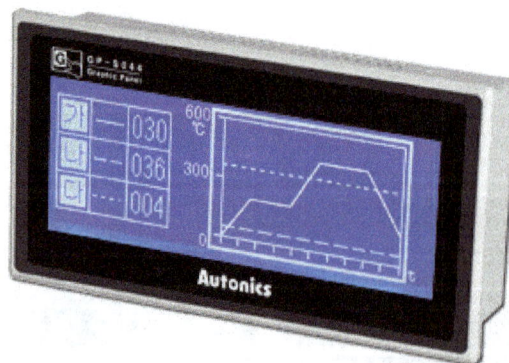

The actual physical properties of an HMI vary from model to model and among manufacturers. It is important that one makes the appropriate selection. An HMI that is located in a water plant might have various water seals around its perimeter, as opposed to an HMI that is located in a pharmaceutical warehouse. The actual size of an HMI is also a key physical property that will vary, because not all applications require a large, high-resolution monitor. Some applications may only require a small, black and white touch screen monitor. When it comes to selecting an HMI, the physical properties are extremely important because one must take into consideration the operating environment, and what safety measures the HMI needs to protect itself. Also, a specific size may be needed due to space limitations. Lastly, physical properties include the processor and memory of the HMI. It is important to make sure that the processor and memory capabilities are sufficient enough to control a system.

## Usage of HMI

HMI technology is used by almost all industrial organizations, as well as a wide range of other companies, to interact with their machines and optimize their industrial processes.

Industries using HMI include:

- Energy
- Food and beverage
- Manufacturing
- Oil and gas
- Power
- Recycling
- Transportation
- Water and wastewater.

The most common roles that interact with HMIs are operators, system integrators, and engineers, particularly control system engineers. HMIs are essential resources for these professionals, who use them to review and monitor processes, diagnose problems, and visualize data.

## Common uses of HMI

HMIs communicate with Programmable Logic Controllers (PLCs) and input/output sensors to get and display information for users to view. HMI screens can be used for a single function, like monitoring and tracking, or for performing more sophisticated operations, like switching machines off or increasing production speed, depending on how they are implemented.

HMIs are used to optimize an industrial process by digitizing and centralizing data for a viewer. By leveraging HMI, operators can see important information displayed in graphs, charts, or digital dashboards, view and manage alarms, and connect with SCADA and MES systems, all through one console.

Previously, operators would need to walk the floor constantly to review mechanical progress and record it on a piece of paper or a whiteboard. By allowing PLCs to communicate real-time information straight to an HMI display, HMI technology eliminates the need for this outdated practice and thereby reduces many costly problems caused by lack of information or human error.

## Developing Trends in HMI Technology

In the past decade, changing operational and business needs have instigated interesting developments in HMI technology. Now, it's becoming more common to see evolved forms of HMI such as high-performance HMIs, touch screens, and mobile devices, along with more traditional models. These modernized interfaces are creating more opportunities for equipment interaction and analysis.

## High-performance HMIs

Operators and users are increasingly moving toward high-performance HMI, a method of HMI design that helps ensure fast, effective interaction. By only drawing attention to the most necessary or critical indicators on the interface, this design technique helps the viewer to see and respond to problems more efficiently, as well as make better-informed decisions. Indicators on high-performance HMI are simple, clean, and purposely cleared of any extraneous graphics or controls. Other design elements, like color, size, and placement, are used with discretion to optimize the user experience.

## Touch Screens and Mobile Devices

Touch screens and mobile HMI are two examples of technological advances that have emerged with the advent of smartphones. Instead of buttons and switches, modernized HMIs allow operators to tap or touch the physical screen to access controls. Touch screens are especially important when used with mobile HMI, which is either deployed through web-based HMI/SCADA or via an application. Mobile HMI offers a variety of advantages to operators, including instant access to HMI information and remote monitoring.

## Remote Monitoring

Mobile-friendly remote monitoring allows greater flexibility and accessibility for operators and managers alike. With this feature, an offsite control system engineer can, for example, confirm the temperature of a warehouse on a portable device, eliminating the need for onsite supervision after working hours. Soon, checking in on a process on your factory floor while being miles away from the facility won't seem like anything out of the ordinary.

## Edge-of-network and Cloud HMIs

Edge-of-network HMIs are also in high demand because they allow operators to access data and visualization from field devices. Additionally, it is becoming more common to send data from local HMIs to the cloud, where it can be accessed and analyzed remotely, while keeping control capabilities local.

## Peering into the Future of HMI

On the horizon, leading engineers are even exploring ways to implement Augmented Reality (AR) and Virtual Reality (VR) to visualize manufacturing functions.

As data takes on an increasingly essential role in manufacturing, the future looks very bright for HMI. This technology may have come a long way, but its potential for growth remains virtually limit less.

## References

- Ouellette, Robert (1983), Automation Impacts on Industry, Ann Arbor, MI, USA: Ann Arbor Science Publishers, ISBN 978-0-250-40609-8.

- An-overview-on-industrial-automation: elprocus.com, Retrieved 11June 2018

- What-is-industrial-automation: electricaltechnology.org, Retrieved 30 April 2018

- Autor, David H. (2015). "Why Are There Still So Many Jobs? The History and Future of Workplace Automation". Journal of Economic Perspectives. 29 (3): 3. doi:10.1257/jep.29.3.3. Retrieved 16 January 2018

- How-process-control-loop-works-in-24: instrumentationtoolbox.com, Retrieved 20 April 2018

- What-is-plc-programmable-logic-controller: unitronicsplc.com, Retrieved 08 July 2018

- Bennett, S. (1993). A History of Control Engineering 1930-1955. London: Peter Peregrinus Ltd. On behalf of the Institution of Electrical Engineers. ISBN 0-86341-280-7

- Engineering-essentials-what-programmable-logic-controller: machinedesign.com, Retrieved 31 March 2018

- Programmable-logic-controllers-and-types-of-plcs: elprocus.com, Retrieved 21 May 2018

# 3

# Automation in Product Design

Automation refers to the technology by which processes are performed without human assistance. This chapter has been carefully written to provide an easy understanding of the varied aspects of automation in product design, such as computer-aided design, boundary representation, computer-aided industrial design, architectural rendering, assembly modeling, 3D modeling, etc.

## Product Design

Product design is the process of identifying a market opportunity, clearly defining the problem, developing a proper solution for that problem and validating the solution with real users.

Design thinking is a method for the practical resolution of problems. Originally coined by David Kelley and Tim Brown of IDEO, design thinking has become a popular approach to creating products. This approach encapsulates methods and ideas of human-centered design into a single unified concept. According to Tim Brown:

> "Design thinking is a human-centered approach to innovation that draws from the designer's toolkit to integrate the needs of people, the possibilities of technology and the requirements for business success".

Good designers have always applied design thinking to product design (whether physical or digital) because it's focused on end-to-end product development, and not just the "design phase" part.

Finding a solution to a problem includes the following five phases:

- Empathize

Learn about the people for whom you are designing. Conduct research to develop a deeper understanding of your users.DefineCreate a point of view that is based on user needs and insights.

- Define

Create a point of view that is based on user needs and insights.

- Ideate

Brainstorm and come up with as many creative solutions as possible. Generate a range of potential solutions by giving yourself and your team total freedom.

- Prototype

Build a prototype (or series of prototypes) to test your hypothesis. Creating a prototype lets designer see if they're on the right track, and it often sparks different ideas that you wouldn't have come up with otherwise.

- Test

Return to your users for feedback.

## Product Design Process

Every design team may follow a different process for product design and development. One process, outlined by Koberg and Bagnell, describes how to turn design ideas into products. The process flows from problem identification to brainstorming ideas, prototype creation and eventually creating the product. This is followed up the formal manufacture of the product and a critical evaluation to identify any improvements that may be needed.

This method includes three stages. The later two may need to be looked at repeatedly during the process.

## Analysis

At the beginning of the process there needs to be extensive research involving concrete facts and figures. This data then feeds into possible solutions to the problem at hand, and the best way to achieve these solutions. Formally, two stages are involved here:

- Accept Situation – The designers commit to the project and identifying a solution. Available resources are consolidated to reach this goal most efficiently.

- Analyze – The team now collectively begins research to collect all relevant data to help reach a solution.

## Concept

Once the problem and potential solutions are narrowed, the final solution is identified and conceptualized in detail. This includes working out adherence to standards and how closely the visualized solution meets identified customer needs. One basic stage here is:

- Define – Here, the team identifies the key issue or issues. Using the problem conditions as objectives and constraints as parameters within which to operate, the team narrows down the information.

## Synthesis

At this stage, the solutions are turned into ideas and the best ones are highlighted. These ideas of design turn into prototypes on which actual products will be based. This stage can be broken down into 4 steps:

- Ideate – Different ideas and solutions are brainstormed here. The best idea bank is created when there is no bias or judgment towards ideas presented.

- Select – The ideas brainstormed are narrowed down to a few which can give the best results. Plans for production can now be created.

- Implement – A prototype can now be created and the plan becomes a product.

- Evaluate – In the final stage, the prototype should be tested and any tweaks necessary should be made. If the prototype does now perform as anticipated, further ideas may need to be brainstormed.

## Product Design Stages

Detailed stages can be followed in a systematic manner to design successful products. These stages includes the following.

## The Design Brief

A statement of intent, the design brief states the problem to be addressed. It serves as a starting point from where the design team can orient themselves. By itself however, it does not offer sufficient information with which to begin the actual design process.

## The Product Design Specification (PDS)

A vitally important but often overlooked and misunderstood stage, the PDS document lists the problem in detail. Before working on producing a solution, there needs to be a deep understanding of the actual problem identified. This document should be designed after conversations with the customer and an analysis of the market and competitors. The design team should refer back to it often for correct orientation at later stages.

## The Concept Design

With the PDS document as a guide, the design team will now begin to outline a solution. At this stage, the design is largely conceptual, with a framework of key components in place with details to a later stage. The details included at this stage will depend on the type of product being designed. It is important to understand both upstream and down-stream concerns relating to the product at this point. These may include activities such as manufacturing, sales and production costs among other things. This early understanding of the value chain will help eliminate or reduce rework and multiple iterations.

In this stage, concept generation and evaluation are both a vital consideration. Multiple concepts, each fulfilling the product requirements previously identified are identified and then evaluated to decide the best way forward.

## The Concept Generation

At this point, a design team may involve a larger audience to help brainstorm the details of concepts drawn up in the previous stage. A group that includes various expertise may end up being the most successful in terms of creative ideas and solutions. It is pertinent to encourage all ideas to be voiced as this increase the chances of innovation.

## The Concept Evaluation

With a number of potential concepts in hand, a suitable design now needs to be chosen that fulfills the product design specifications previously generated. This document should serve as a basis for final design decisions. Again, a multi skilled team should be involved here so that all angles of the chosen design can be evaluated. The concept that is closest in solving the problem identified and fulfills the most design requirements will now be developed in detail.

## The Detailed Design

At this point, the final concept has been chosen and most obvious kinks have been worked out. The concept is now designed in detail with the necessary dimensions and specifications. At this stage, it may be important to produce one of more prototypes to test the product in close to real scenarios. It becomes vital for the design team to work in close cooperation with other units such as manufacturing and logistics to ensure the practical aspects of production and supply.

## Eliminating Design Iterations

Although traditionally sequential, multiple iterations within these stages can be reduced by asking the following questions:

- Manufacturing – Can we make the product at our existing facility.
- Sales – Are we able to produce what the customer wants.
- Purchasing – Do we have required parts available or do they need to be ordered.
- Cost – What will the design cost us to make.
- Transport – is the product sized for available transportation methods, Will there be any special transportation needs.
- Disposal – How will the product be disposed of at the end of its life.

## Product Design Types

Two basic categories encompass most product designs.

- Demand – Pull Innovation

Demand – Pull happens when a product design can directly take advantage of an opportunity in the market. A new design works towards solving an existing design issue. This happens either through a new product or a variation of an existing product.

- Invention – Push Innovation

This innovation occurs with an advancement in technology or intelligence. This is driven through research or a creative new product design.

## Factors Affecting Product Design

### Cost

One major factor that affects product design is the cost of production including material costs and labor costs. These in turn affect the pricing strategy, which needs to be in line with what the customer is prepared to pay for it.

### Ergonomics

The product needs to be user friendly and afford convenience in its function. Using ergonomic measurements, minor or major changes may need to be made to product design to meet essential requirements.

### Materials

Whether the requisite materials are available easily is an important consideration in product design. In addition, an eye needs to be kept on new developments in materials and technology.

### Customer Requirements

One major and obvious influence on the design on the product is the customer and their requirements. It is vital to capture customer feedback on any prototype as well as during the planning and conceptual stages. Even a technologically advanced and exciting feature may need to be removed if it causes dislike or negative feelings in an end user.

### Company Identity

The company's identity is a point of pride and as a matter of course, a product's very design or color schemes and features may be determined by this identity. The logo may need to be featured in a specific manner or subtle or overt features of the company identity may need to be built into the design.

## Aesthetics

The product may need to appear stylish or of a certain shape. This form may end up determining the technology that it built into the product. This may in turn also affect the manufacturing process that needs to be followed.

## Fashion

The current fashion and trends may also affect a certain product's design. Customers will want the most updated options and this need to be considered during product design.

## Culture

If a product is for a certain market with its own individual culture, this needs to be kept in mind during product design. A product acceptable in one culture may end up being offensive or not desirable in another one.

## Functions

How many problems is the product trying to solve? The number of uses and functions a product has will impact its design.

## Environment

Another consideration to product design is its impact on the environment. The average customer these days may be more discerning and concerned about the environment than before. Things to consider here may include whether the materials used are recyclable, how the product will be disposed of at the end of its life or how the packaging can be disposed of.

## Consideration in Product Design

Product design is a complex process, since all the relevant stakeholders have different requirements from the product. An example of conflicting needs that will require attention during product design are mentioned below.

## Economic Viability

The manufacturer will want the product to be created at the lowest cost possible, in order to maximize profit and ensure sales. A prohibitively expensive product will have higher price tag and may drive away customers. Often, this may mean a product redesign or a compromise on quality.

## Price, Appearance and Prestige Value

The customer will always want a well presented product with a functional yet aesthetically appealing design. They will also want it to be priced reasonably. The appearance may not always be vital to function, but if there are multiple nearly similar products in the market, the look of the product may become the deciding factor.

## Functionality

There needs to be equal focus on the functionality of the product or how well it performs. This is a given as the product foremost needs to perform as it claims to. The end user may purchase for the external appearance. But long term satisfaction and repeat usage will only occur if the product performs at an optimal level.

## Maintenance

Product designers, manufacturers and maintenance workers may all favor a modular construction for a product. The more easily different parts can be worked on individually, the more versatility the product offers. A re-design effort may only need to focus on changing certain parts rather than the whole, the manufacturer can easily tweak elements without changing entire production processes and maintenance workers may not need to disassemble everything, thereby reducing repair time and effort.

## Examples of Successful Product Design

### The Apple iPhone

Apple is consistently ranked as one of the most innovative companies in the world. Though not always successful initially, Apple has managed to create unique products with superior designs that have great appeal with end users. The Apple iPhone revolutionized the cell phone market with its innovative features, streamlined design and an entire supporting universe through the app store. Though it was not the pioneer in smart phones, Apple is extremely successful because it created a beautiful product that gives a superior user experience to a consumer.

### The Porsche Cayenne

A respected and coveted sports car maker, Porsche entered the SUV market a few years ago. The sports utility vehicle is thought to be boring and the category has broad generic definitions of utilitarian. Porsche, through the Cayenne, endeavored to bring together the rational and the emotional. Through focused product design, the company managed to build an SUV that has all the necessary features of this category of car but with the driving experience of any other Porsche car. This resulted in successful sales and made this car a breakthrough product into a new market for a premium brand.

### 6Wunder Kinder

Launched by a young, innovative start-up, this company launched a cloud based, cross platform productivity application called Wunderlist in 2010. It is now one of the most popular apps of its kind and boasts millions of users. It successfully launched across five different platforms in a brief span of time. Data is seamlessly updated across multiple devices and the design is simple but effective.

When designing a new product or re-designing an existing, it is pertinent for the company to clearly identify what problem it is attempting to solve through this new product. It is vital to involve a multi skilled team in order to ensure a critical view of all ideas and in fact to offer a wider and more innovative pool of ideas to choose from. It is also very important to consider the customer and their requirements and desires from the very beginning of the product till the very end. It is also a good idea to not get emotionally attached to a design. This can make the team lose focus of what is to be done in order to create and successful and lasting product. Instead, a critical and analytical view of the process should be taken, with any changes that are necessary being made.

With this approach and mindset, creating successful product designs will become a more systematic process and the resulting products will leave a long term impact on the consumers.

## Computer Aided Design

Computer-aided design (CAD) is a computer technology that designs a product and documents the design's process. CAD may facilitate the manufacturing process by transferring detailed diagrams of a product's materials, processes, tolerances and dimensions with specific conventions for the product in question. It can be used to produce either two-dimensional or three-dimensional diagrams, which can then when rotated to be viewed from any angle, even from the inside looking out. A special printer or plotter is usually required for printing professional design renderings.

The concept of designing geometric shapes for objects is very similar to CAD. It is called computer-aided geometric design (CAGD).

CAD is also known as computer-aided design and drafting (CADD).

CAD is used as follows:

- To produce detailed engineering designs through 3-D and 2-D drawings of the physical components of manufactured products.

- To create conceptual design, product layout, strength and dynamic analysis of assembly and the manufacturing processes themselves.

- To prepare environmental impact reports, in which computer-aided designs are used in photographs to produce a rendering of the appearance when the new structures are built.

CAD systems exist today for all of the major computer platforms, including Windows, Linux, Unix and Mac OS X. The user interface generally centers around a computer mouse, but a pen and digitizing graphic tablet can also be used. View manipulation can be accomplished with a space mouse (or space ball). Some systems allow stereoscopic glasses for viewing 3-D models.

Most U.S. universities no longer require classes for producing hand drawings using protractors and compasses. Instead, there are many classes on different types of CAD software. Because hardware and software costs are decreasing, universities and manufacturers now train students how to use these high-level tools. These tools have also modified design work flows to make them more efficient, lowering these training costs even further.

## Advantages and Disadvantages

Modeling with CAD systems offers a number of advantages over traditional drafting methods that use rulers, squares, and compasses. For example, designs can be altered without erasing and redrawing. CAD systems also offer "zoom" features analogous to a camera lens, whereby a designer can magnify certain elements of a model to facilitate inspection. Computer models are typically three dimensional and can be rotated on any axis, much as one could rotate an actual three dimensional model in one's hand, enabling the designer to gain a fuller sense of the object. CAD systems also lend themselves to modeling cutaway drawings, in which the internal shape of a part is revealed, and to illustrating the spatial relationships among a system of parts.

To understand CAD it is also useful to understand what CAD cannot do. CAD systems have no means of comprehending real-world concepts, such as the nature of the object being designed or the function that object will serve. CAD systems function by their capacity to codify geometrical concepts. Thus the design process using CAD involves transferring a designer's idea into a formal geometrical model. Efforts to develop computer-based "artificial intelligence" (AI) have not yet succeeded in penetrating beyond the mechanical—represented by geometrical (rule-based) modeling.

Other limitations to CAD are being addressed by research and development in the field of expert systems. This field is derived from research done in AI. One example of an expert system involves incorporating information about the nature of materials—their weight, tensile strength, flexibility, and so on—into CAD software. By including this and other information, the CAD system could then "know" what an expert engineer knows when that engineer creates a design. The system could then mimic the engineer's thought pattern and actually "create" more of the design. Expert systems might involve the implementation of more abstract principles, such as the nature of gravity and friction, or the function and relation of commonly used parts, such as levers or nuts and bolts. Expert systems might also come to change the way data are stored and retrieved in CAD/CAM systems, supplanting the hierarchical system with one that offers greater flexibility. Such futuristic concepts, however, are all highly dependent on our abilities to analyze human decision processes and to translate these into mechanical equivalents if possible.

One of the key areas of development in CAD technologies is the simulation of performance. Among the most common types of simulation are testing for response to stress and modeling the process by which a part might be manufactured or the dynamic relationships among a system of parts. In stress tests, model surfaces are shown by a grid or mesh, that distort as the part comes under simulated physical or thermal stress. Dynamics tests function as a complement or substitute for building working prototypes. The ease with which a part's specifications can be changed facilitates the development of optimal dynamic efficiencies, both as regards the functioning of a system of parts and the manufacture of any given part. Simulation is also used in electronic design automation, in which simulated flow of current through a circuit enables the rapid testing of various component configurations.

The processes of design and manufacture are, in some sense, conceptually separable. Yet the design process must be undertaken with an understanding of the nature of the production process. It is necessary, for example, for a designer to know the properties of the materials with which the part might be built, the various techniques by which the part might be shaped, and the scale of production that is economically viable. The conceptual overlap between design and manufacture is suggestive of the potential benefits of CAD and CAM and the reason they are generally considered together as a system.

## Types

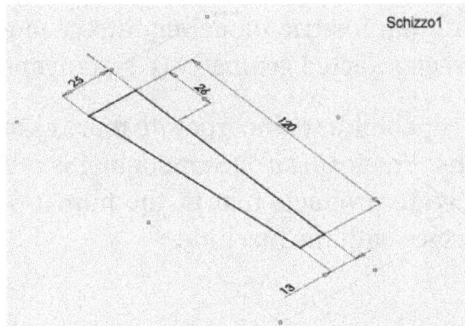

There are several different types of CAD, each requiring the operator to think differently about how to use them and design their virtual components in a different manner for each.

There are many producers of the lower-end 2D systems, including a number of free and open source programs. These provide an approach to the drawing process without all the fuss over scale and placement on the drawing sheet that accompanied hand drafting since these can be adjusted as required during the creation of the final draft.

3D wireframe is basically an extension of 2D drafting (not often used today). Each line has to be manually inserted into the drawing. The final product has no mass properties associated with it and cannot have features directly added to it, such as holes. The operator approaches these in a similar fashion to the 2D systems, although many 3D systems allow using the wireframe model to make the final engineering drawing views.

3D "dumb" solids are created in a way analogous to manipulations of real-world objects (not often used today). Basic three-dimensional geometric forms (prisms, cylinders, spheres, and so on) have solid volumes added or subtracted from them as if assembling or cutting real-world objects. Two-dimensional projected views can easily be generated from the models. Basic 3D solids don't usually include tools to easily allow motion of components, set limits to their motion, or identify interference between components.

## Types of 3D Solid Modeling

Parametric modeling allows the operator to use what is referred to as "design intent". The objects and features created are modifiable. Any future modifications can be made by changing how the original part was created. If a feature was intended to be located from the center of the part, the operator should locate it from the center of the model. The feature could be located using any geometric object already available in the part, but this random placement would defeat the design intent. If the operator designs the part as it functions the parametric modeler is able to make changes to the part while maintaining geometric and functional relationships.

Direct or Explicit modeling provide the ability to edit geometry without a history tree. With direct modeling, once a sketch is used to create geometry the sketch is incorporated into the new geometry and the designer just modifies the geometry without needing the original sketch. As with parametric modeling, direct modeling has the ability to include relationships between selected geometry (e.g., tangency, concentricity).

Top end systems offer the capabilities to incorporate more organic, aesthetics and ergonomic features into designs. Freeform surface modeling is often combined with solids to allow the designer to create products that fit the human form and visual requirements as well as they interface with the machine.

## Technology

A CAD model of a computer mouse

Originally software for Computer-Aided Design systems was developed with computer languages such as Fortran, ALGOL but with the advancement of object-oriented programming methods this has radically changed. Typical modern parametric feature-based modeler and freeform surface systems are built around a number of key C modules with their own APIs. A CAD system can be seen as built up from the interaction of a graphical user interface (GUI) with NURBS geometry or boundary representation (B-rep) data via a geometric modeling kernel. A geometry constraint engine may also be employed to manage the associative relationships between geometry, such as wireframe geometry in a sketch or components in an assembly.

Unexpected capabilities of these associative relationships have led to a new form of prototyping called digital prototyping. In contrast to physical prototypes, which entail manufacturing time in the design. That said, CAD models can be generated by a computer after the physical prototype has been scanned using an industrial CT scanning machine. Depending on the nature of the business, digital or physical prototypes can be initially chosen according to specific needs.

Today, CAD systems exist for all the major platforms (Windows, Linux, UNIX and Mac OS X); some packages support multiple platforms.

Right now, no special hardware is required for most CAD software. However, some CAD systems can do graphically and computationally intensive tasks, so a modern graphics card, high speed (and possibly multiple) CPUs and large amounts of RAM may be recommended.

The human-machine interface is generally via a computer mouse but can also be via a pen and digitizing graphics tablet. Manipulation of the view of the model on the screen is also sometimes done with the use of a Spacemouse/SpaceBall. Some systems also support stereoscopic glasses for viewing the 3D model.Technologies which in the past were limited to larger installations or specialist applications have become available to a wide group of users. These include the CAVE or HMDs and interactive devices like motion-sensing technology.

## Computer Aided Industrial Design

Computer-aided industrial design (CAID) is simply the use of computerized software in the industrial design process. As opposed to traditional manual drafting, CAID is an automated process that greatly increases the efficiency of design alterations, concept testing and general optimizations. CAID grants designers creative freedom, however it is common to follow a simple methodology: a designer will create a sketch using a stylus, following which they will generate curves from the sketch and in turn generate surfaces from the curves.

## Importance of Computer-aided Industrial Design

CAID software allows a designer to create a 3D model prior to the manufacturing of the product itself. The 3D model can be saved in a format that can be read by a rapid prototyping machine which will then create a real-life model of the product. These computerized steps speed up the creation process, as the designer has more time to focus on the technical aspects of the design rather than sketching and modeling manually. This allows for a better product proposal in a shorter amount of time.

## Computer-aided Modeling Technology

In the area of CAID technology, computer-aided modeling technology is mainly reflected in the shape of the freeform surface design and sketch design. In the free-form surface design, the product appearance freeform surface design study is an important content of the CAID. And the surface feature design is an important development in the design of free-form surfaces. The surface feature design includes three parts, namely basic surface, mobile features and collusion graphics.

## The High-tech of CAID

Currently, the market began to slip from the emerging technologies in the high-tech of CAID, such as virtual reality, genetic algorithms and so on. But how to use these technologies well in CAID field, this would be carried out with some of the traditional technologies effectively, thereby to approach some CAID related research. The collaborative, parallel design is now one of the main development directions of this technology. As for the industrial design, the sense of product design is very important, a detailed study of the product functions, principles, shape must be carried out. Many scholars observe on a variety of contents in some parallel environment in different angles of the technology and deep into it to understand and explore. But the starting point of other scholars is from the CSCW perspective, they have carried out a detailed analysis of the collaborative design and studied the use of this design, explored the model of engineering and industrial design work.

## Intelligent Technology of CAID

Currently, the Intelligent CAD has witnessed a considerable degree of development. Integrated Intelligent Design System (I2CAD) provides an integral computer support to the design during the whole process. As it relates to the creative thinking as well as the frequent human-computer interaction, industrial design, in particular, needs auxiliary artificial intelligence technology. In the industrial design process of creative design thinking, the translation of the designer's conception fast into sketch is a fairly complex process behavior, and this process is known as the stage of concept function. Many scholars make researches on the stage of concept function from the designer's creative design thinking, design knowledge representation, and put forward their own views with a combination of sketch design. Design grammar, it is the formal description method by refining and abstracting the elements of shape, color, and shape of the object and its generation principle from the angle of design methodology. It is one of the foundations of intelligent design system. At present, the design grammar of the industrial design community scholars mainly includes the pose grammar, color grammar, shape grammar and modeling conversion grammar.

## Key Technologies to be Solved

- Research of modern design methodology. Based on the development direction of the modern industrial design, makes research on the qualitative design process and the design method with the similar accurate method to lay the theoretical foundation of CAID from the design object itself.

- Research of innovative design technology. Follow the principles and norms obeyed by the design process of research design thinking process and computer support; explore a wide range of innovative techniques, study innovative design principles, methods and techniques in depth.

## Influence of the Computer to the Concept and Method of Industrial Design

Due to the development of computer software, the product has made a great progress in the design of the degrees of freedom. In the traditional design, the expression of hyperboloid and the free-form surface is very troublesome, and it often needs to produce a solid model to express clearly. It is also a difficult thing to change model again into engineering drawing. Therefore, in the design, the designer always avoids the use of free-form surfaces, which makes the design conservative. Today, with the use of computer to generate data model, all these difficulties are gone, and the relationship between design and manufacturing is closer. The use of computers makes us change the design criteria. Traditional design puts high demands on the effect of expression, it often takes that whether the drawing production is sophisticated, the line is light, and the color is uniform as an important criterion of evaluation. However, this criterion loses its meaning due to the computer-precise data and sophisticated output. Meanwhile, we put the evaluation criteria on the evaluation of the merits of the design. Moreover, the Computer-aided design has shortened the product development cycle. On one hand, it increases the efficiency of the work; on the other hand, it eliminates many steps of the traditional design performance. Especially on the program modifications and adjustments, it is very convenient to modify because the computer retains the whole process of design.

## Product Modeling CAID

Traditional mechanical design and manufacturing is cumbersome and difficult to modify because of the use of artificial mapping, it generally only draws view. It is very difficult to draw a perspective view for complex models, designers can only base on its plan to imagine the finished model after the three-dimensional model, which is very difficult to design and production. However, computer assisted cartography changes all these. For example, a common base, the computer maps out its three-view, the shaft side maps, and cross-sectional view only in5 min, and computer automatically marks all sizes. when you want to change a size, all views are automatically amended accordingly axis the angle of the side of the map, cross-sectional view of the cutting position can be adjusted and can be displayed to direct three-dimensional effect. Simulating products work environment, rendering module, assigning different materials, designing products appearance, drawing idea sketches and design effect diagram, mimic motion effects, analyzing movement interference, with this end, we can watch last made to improve efficiency and to avoid losses caused by ill-considered design at the design stage.

## Conceptual Effect Drawing Stage of Modeling Design

In the early stages of the modeling creative design, industrial design promotes hand-painted, for hand-painted is the most natural way for designers to capture the inspiration, as long as there is a piece of paper, a pen, they can record their inspirations at anytime, anywhere. Hand drawing even should be a means of inspiration record because inspiration cannot be controlled. Only accumulating in daily life, will you be able to come in handy in a lot of materials. When you combine these materials with your personal style as well as product design orientation, you can go back to work in front of the computer.

## Characteristics and Application

### The Main Features

CAID technology has unparalleled advantages than the traditional industrial design, industrial designers can free to express a creative idea to display their talents if master it. It can enhance the quality of the overall product design, strengthen the competitiveness of the product market, and has the following characteristics: high-quality three dimensional space software system set a three-dimensional solid modeling, static coloring, complex lighting model, and the multimedia animation in one, vivid image. It ensures the high quality of the design through advanced design tools. Flexibility with such high-tech tools for creative design directly on the system, using 3D solid modeling techniques for geometric modeling of objects, such as a color design, material editing, form, texture depicting real-time rotation transformation, rapid real image generate output, many different styles, program evaluation and testing. It can easily be modified until satisfied. The system is to optimize the design.

### The Application of High-tech Research in CAID

Currently, these emerging technologies of virtual reality, neural networks, genetic algorithms and parallel design, collaborative design method are the hot spots of the majority of scholars. The introduction of these technologies into CAID field, combination of traditional optimization design, fuzzy technology, intelligent technology to CAID study also gradually win the attention of scholars.

Parallel design, collaborative design is one of the trends of modern design. In the field of industrial design, especially product design, it is necessary to study the parallel and collaborative design mechanism of product features, principle, and layout. From Concurrent engineering point of view, some scholars have explored deeply into the parallel design process and design environment for concurrent engineering design technology.

## Positioning and its Implementation

Analysis through in-depth investigation of a number of factories, refrigerators, machine tools, and other typical product development process, various design techniques can be seen in the application of the entire product development process and the role of various designers. Based on accurate position, they can develop CAID system development strategy and implementation method. Computer-aided industrial design (CAIDS) is developed in accordance with the CAID technical principles and methods of computer-aided design software system.

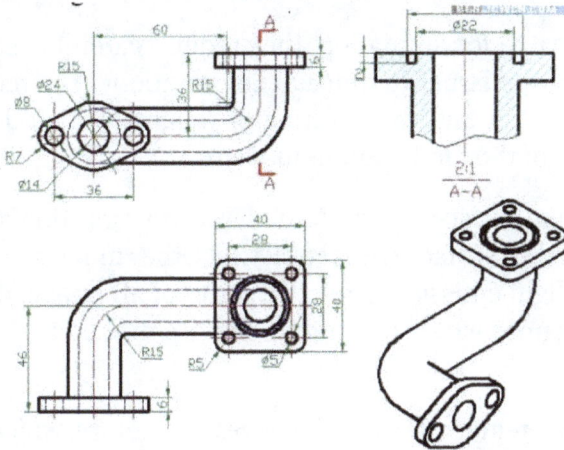

Computer-aided design software system

It should be in accordance with the thinking of system engineering, completely take industrial design theory and methods as guidance for the smart innovative product development and design system. it first does form design and human-computer design in the machine; then imports the product model into design platform, including design of color, decoration, material, etc.; at last, it comes to the conclusion of the product modeling program modeling program expression including renderings of product modeling, design, evaluation, and engineering geometric model. The product designed through the system has the features of good shape and beautiful color, pleasant, high quality, efficient, animation and others.

## Application Examples

It can effectively form a product family or a different design for the evaluation and selection. Currently it mainly uses algorithms include adaptive neural network and morphological differences in residual algorithm to control the generation of the new design, and the method of adaptive neural network is applied boarder.

## 3D Modeling

A 3D model is a mathematical representation of something three-dimensional.

3D models are used to portray real-world and conceptual visuals for art, entertainment, simulation and drafting and are integral to many different industries, including virtual reality, video games, 3D printing, marketing, TV and motion pictures, scientific and medical imaging and computer-aided design and manufacturing CAD/CAM.

Some 3D models are constructed from NURBS (non-uniform rational b-spline), smooth shapes defined by bezel curves, which are relatively computationally complex. The typical base of a the model is a 3D mesh; the structural build consists of polygons.

When models are created for animation, they require careful construction because the polygon layout can create issues in unusual deformations. The models also require the construction of a skeleton and the painting of weights, which define the texture and polygon deformation of the model under movement.

Some 3D models define surfaces through shaders, programs that mathematically define color, lightplay and other surface characteristics. Other models define color, secularity, surface texture, and light emission through a series of 2D image files called maps, especially those used in games where raster graphics are needed to deliver real-time frame rates.

A more recent development in 3D modeling is reality capture, which uses remote sensing technology such as Lidar to capture complex forms quickly and accurately. Reality capture may be used in combination with 3D printing for an end-to-end process known as reality computing.

3D modeling is used in a wide range of fields, including engineering, architecture, entertainment, film, special effects, game development, and commercial advertising.

A popular example of 3D technology is its use in major motion pictures. Just think of the scenery in the film Avatar, the 2009 movie from director James Cameron. The film helped to transform the 3D industry when it used many of the concepts of 3D modeling to create the movie's planet Pandora.

3D modeling is fun but difficult. Unlike many graphic fields, 3D modeling requires a substantial learning curve and sophisticated software. Beginners in 3D can be put off by the time needed to master 3D modeling, but, with patience, they can be turning out animations, structural renderings, and video game graphics in no time. It is likely that the software you choose to use comes with a wealth of online tutorials or instructional classes. Take advantage of these resources to come up to speed with the software and 3D modeling.

There are 4 general types of 3D modeling methods, which are listed below.

- Polygonal Modeling: A polygonal model represents points in 3D space connected by line segments to form a polygon mesh. Polygonal mesh files are planar, which means that they are represented by a series of flat facets. Therefore, curves can only be approximated through surface subdivision with a defined resolution. Polygonal meshes are convenient because they are lightweight and visualizations can be rendered quickly.

- Curve Modeling: Another type of modeling that relies on curves to generate surface geometry. Curve modeling can be both parametric (based on geometric and functional relationships) or freeform, and rely on NURBS (non uniform rational B-splines) to describe surface forms. The curves are driven by mathematical equations that are influenced by the designer using weighted control points. Check out our entry on Rhinoceros to learn more about curve modeling.

- Digital Sculpting: This is a relatively new type of 3D modeling where the user interacts with the digital model as you would modeling clay. Users can push, pull, pinch, or twist virtual clay to generate their model. Sculptris is a great example of digital sculpting.

- Code-driven modeling: this is a growing area of modeling where geometry is generated autonomously based on conditions set in place by the designer. To read more about how this works, check out our entry on Autodesk's Deamcatcher . This type of modeling is excellent for 3D printing as it can be used to generate 3D structures that can't be manufactured through any other means.

## Printing

3D modeling can help with both exclusive and mass production. You can create a 3D model for molding or for using with a printer.

## Visualization

After visualization one usually gets one or a few 3D pictures of an object. Designers not only work with the geometry, but with textures, materials, and lighting as well. Images are edited in Photoshop afterwards.

## Stages of Creating a Photorealistic 3D Image

Stage 1: 3D modeling

Creation of the object's geometry according to the pre determined measurements or on a scale. There are plenty of programs and tools for making 3D models, such as Autodesk 3ds Max, Autodesk Maya, Autodesk Revit, Modo, Google SketchUp, Z-Brush etc. If you need a free program, use Blender.

Stage 2: recreating physical qualities of textures and materials to make the image more realistic

All materials have different colors and surfaces. They can also reflect or absorb light. This is why it is necessary to fine-tune their transparency, gloss, smoothness or roughness etc.

Stage 3: choosing light sources

Lighting can be artificial or natural, which influences the look of the modeled object greatly. It is important to determine brightness and depth, and also add shadows. If you do it the right way, the image will be much more realistic.

Stage 4: placing virtual cameras

It is necessary to choose the right angles for the image.

Stage 5: rendering

Or projecting the image onto a 2D surface. The most popular rendering software includes V-Ray, Corona Render, Photo Realistic Render Man (PR Man), Maxwell Render.

Stage 6: postproduction

Or processing of ready images with the help of Adobe Photoshop, Adobe after effects Pro, Adobe Premier Pro, Adobe Photoshop Light room. On this stage, it is possible to add surroundings and additional effects or shades which would make images more atmospheric (for example, depict weather conditions).

A 3D model can be used several time after its completion. Because of this, now there is a market of ready-made models. Using ready models, designers can choose whatever they need for their projects and lower their expenses.

## Model-based Definition

When you look at a 3D model, you can see the shape of a product, but you often don't know all the engineering details required to successfully manufacture the real-world version. To get the complete picture, you need dimensions, surface finish, materials, tolerances, and more.

Traditionally, that non-geometric information appears in separate 2D drawings or documents. Engineers develop the product as a 3D model, and then add the product manufacturing information (PMI), as annotations on those drawings.

But what happens when the model is complex and downstream processes, vendors, partners, etc., can't easily discern the geometry or annotations? Or when the model is updated and all those drawings need to be updated? This problem isn't unusual, and leads to a lot of waste.

That's the problem that model based definition (MBD) seeks to solve.

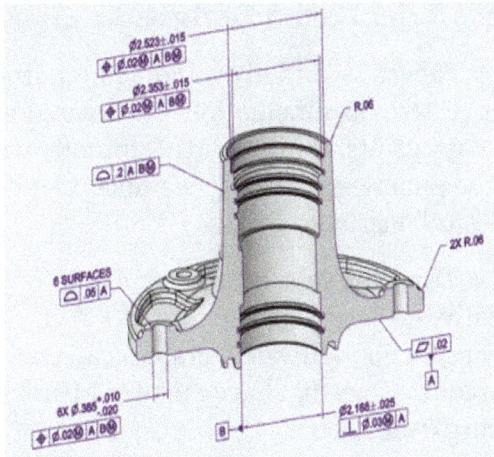

Model Based Definition (MBD) is a mechanical engineering initiative where a 3D model with Product Manufacturing Information (PMI) augments or replaces a 2D engineering drawing as design documentation.

Product and Manufacturing Information (PMI) is composed of non-geometric information that describes a design, including Geometric Dimensioning and Tolerance (GD&T), surface finish, material information and more. Geometric Dimensioning and tolerance (GD&T): is an approach to defining tolerances that uses a symbolic language.

Design documentation refers to the deliverables that engineering releases to down-stream functional organizations such as manufacturing, procurement, suppliers, service and much more.

MBD deliverables are created and consumed with Mechanical Computer Aided Design software or 3D Visualization software. Such deliverables can be managed with Product Data Management software or Product Lifecycle Management software.

### Reasons for Pursuing a Model Based Definition Initiative

There are a few reasons an organization would pursue this initiative, including:

- Contractual Obligations: A number of today's Department of Defense contracts include terms and conditions requiring the delivery of 3D models with PMI. For some, their pursuit of this initiative is primarily driven by contractual obligation.

- Reducing Efforts to Creates and Amend Design Documentation: Theoretically, embedding PMI onto a 3D model should take less time than detailing a complete 2D engineering drawing, promising some benefit from this initiative. Furthermore, 3D models with PMI is often clearer in communicating manufacturing intent, allowing engineers to avoid the creation of amending documentation to clarify 2D engineering drawings.

### Clarifying Points on Model Based Definition

- MBD Initiatives ≠ Paperless Initiatives: Some organizations are pursuing paperless initiatives, where documentation is delivered electronically instead of on physical, hardcopy prints. While an MBD initiative may also be paperless, it does not have to be paperless. MBD deliverables can be printed on paper and distributed in physical, hardcopy forms.

- MBD Initiatives ≠ Drawling less Initiatives: Some organizations are pursing drawing less initiatives, where a 3D model and PMI is delivered instead of a 2D engineering drawing. While MBD deliverables can enable such efforts, it is not required. Some organizations that have pursued MBD initiatives use 3D models and PMI alongside 2D engineering drawings.

# Architectural Rendering

Architecture rendering is a form of drawing that architects use in order to show off the designs that they have come up with. Unlike visualisations; which can be two dimensional or three dimensional, architecture rendering uses two dimensional designs only.

Usually, an architect will put together a two dimensional line drawing that shows the building from a specific perspective. They may make a series of drawings showing different sides and the interior or exterior of the building.

These drawings could be line drawings only, or could be more detailed colored sketches depending on how far through the process of design the architect is.

## Assembly Modeling

Many of the modelers available today may best be classified as geometric modelers. These systems have data structures that have been designed to store and manipulate geometric data of individual parts only. Assembly modelers can be thought of as more advanced geometric modelers where the data structure is extended to allow representation and manipulation of hierarchical relationships and mating conditions that exist between components in an assembly (Zied, 1991). As shown in figure below, the geometric modeler acts as a front end to the assembly modeler. Individual parts may be modeled using the geometric modeler. These models may be analyzed individually at this stage. After each part has been analyzed and optimized, the designer would use the assembly modeler to synthesize an assembly model and analyze the entire system.

Inherent in all assemblies is the notion of hierarchical relationships. This implies a layered or tree-like structure for the assembly, as shown in The tree (also called assembly tree) explodes the overall assembly into subassemblies and parts, as well as illustrates where within the tree structure the various parts and subassemblies are connected or attached. Thus at the topmost level of the hierarchy or depth 0, we have the overall assembly. The next level, i.e. depth 1, shows how the major subassemblies and parts fit into the overall assembly. This process of exploding and detailing continues until all subassemblies, parts and components have been accounted for. With the benefit of such hierarchical relationships, questions such as whether gear 1 belongs to subassembly 1 or subassembly 3 can be addressed.

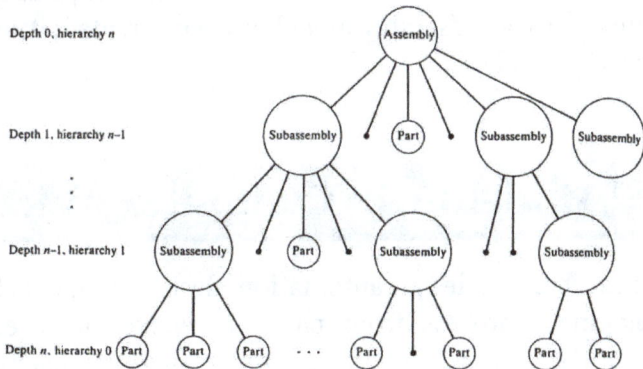

In assembly modeling, individual parts modeled by a geometric modeler are combined together using "merge" commands to form the assembly. Typically a designer would start with a base part (the largest component into which most others fit) and add other components to the base part using merge commands. These merge conditions define spatial relationships (also called mating conditions or assembly constraints) between individual parts. For example, mating conditions specify whether two planar surfaces butt against each other, or if the axis of a hole and shaft are coaxial or otherwise. Mating conditions are interactively specified with great ease in assembly modelers.

Each part has associated with it degrees of freedom or independent movements. Generally there are six degrees of freedom to be considered: three translational and three rotational movements. As parts are assembled using merge commands, their associated degrees of freedom are constrained or restricted. For example, when a nut is assembled onto a hole in a part, the nut loses three translational and two rotational degrees of freedom. Such constraining of the degrees of freedom is essential for fully describing the spatial relationship between mating parts.

Once hierarchical relationships and mating conditions have been described for a set of mating parts, the result obtained is called an assembly model. The difference between assembly and solid modelers in terms of assembly building may be understood by the following example. If a robot consisting of six rigid links that are connected together by some translational (prismatic) and some rotational (revolute) joints is modeled by a solid modeler and assembled together by using mere "move" and "translate" commands, then an analysis program may not be able to use this assembly to decide if a torque applied to link 1 will cause link 4 to move or not. Furthermore, if movement should result, the program may not be able to decide whether the resulting movement would be translational or rotational. If the same robot were built using an assembly modeler, then it could support this analysis.

Assembly modelers can also automatically generate bill of materials (BOM), determine mass properties for the entire assembly, graphically "explode" the assembly to show how the components fit together, list parent child relationships and generate multi-views of the assembly. Once an assembly model has been created, it can be subjected to some sort of engineering analysis.

## Behavioral Modeling in Computer Aided Design

With the advent of mechanical design automation, design engineers began to transfer the product development process from paper to electronic-based models, using the first three generations of technology— initially, two-dimensional drafting, then three-dimensional wireframe modeling, and finally three-dimensional solid modeling. The current 4th generation of mechanical design automation is the most significant to date, and takes advantage of associative, parametric feature based modeling to improve design efficiency and expand the advantages of electronic-based design throughout the engineering process.

However, as product requirements become more volatile and exhaustive and products more highly focused and tailored for additional markets, engineers require a different type of mechanical design automation technology. The approach that addresses these expanded needs has recently been solidified and consolidated in the industry as the 5th generation of Behavioral Modeling.

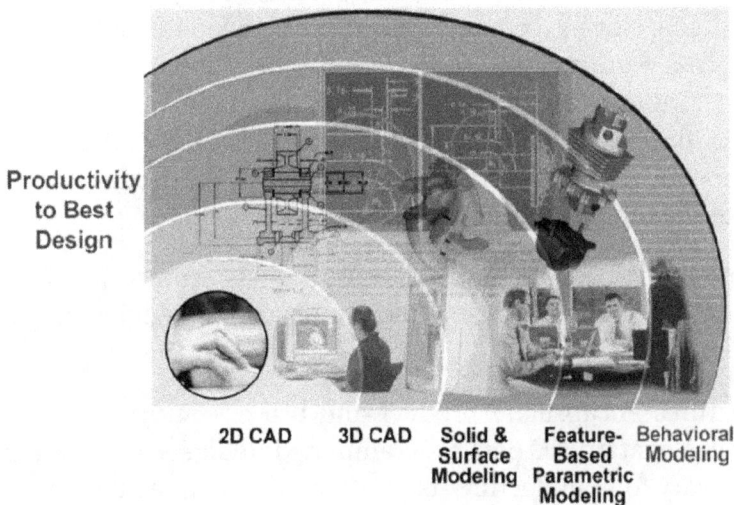

| 2D CAD | 3D CAD | Solid & Surface Modeling | Feature-Based Parametric Modeling | Behavioral Modeling |

The Evolution of Mechanical Design Automation Technology

Behavioral Modeling promotes the creation of well designed products through the synthesis of requirements, desired functional behavior, design context and geometry through an open, extensible environment. Built on the solid foundation of an associative, feature-based, parametric modeling kernel, Behavioral Modeling raises the level of mechanical design automation from simple geometry creation to fully engineering a solution.

## Cornerstones of Behavioral Modeling

The power of Behavioral Modeling eminates from three forces: smart models, objective-driven design capabilities and an open extensible environment. Smart models capture design and process intelligence and the range of engineering specifications required to define a product. Objective driven design capabilities optimize each product design to satisfy multiple objectives and changing market needs. An open extensible environment enables organizations to integrate diverse tools across unique engineering processes.

## Smart Models

Smart models are intelligent designs that contain all the specification and process information they need to adapt to their environment. Because smart models are "aware" of their context and purpose, they enable organizations to develop more innovative, differentiated, and customer responsive products.

## Capturing Design Information in Features

Smart models capture this intelligence within features. Originally, feature-based modeling, which was first introduced to the market with Pro/ENGINEER, only described the geometry and form of a design, such as protrusions, holes, rounds and blends. A Behavioral Modeling strategy advances feature-based modeling to accommodate a set of adaptive process features that go beyond the traditional core geometric features. These next-generation extended features accommodate a variety of information that further specifies the intent and performance of the design. Like the traditional geometric features, these other features are an integral part of the product model and perform like any other design object in the system.

There are two distinct categories of adaptive process features. Application features describe process information. Behavioral features contain engineering and functional specifications.

Application features encapsulate product and process information. For example, a pocket feature can contain NC part programming intelligence, including the tools and tool paths necessary to manufacture itself. Routed-system features, such as cable segments and spools, simplify cable routing to point and click, because they hold all the necessary information to capture connectivity data, and to self-determine appropriate paths to avoid violating minimum bend radii.

Behavioral features contain information about design specifications, such as desired weights, angles of reflection, mass properties, or other measurements. They can also include spatial allocation information, including external static envelopes (shrink wraps) and working envelopes (swept solids/movement evaluations). Moreover, when product definitions require the use of a customer's proprietary design application (for instance, one that designs turbine blade profiles), a behavioral feature can reference that external application as part of the product model definition.

Behavioral features can also define the way that components in an assembly connect, using welds or pin, ball, and slider joints. When behavioral features consist of assembly connection information, including any assembly constraints, the assembly design process automatically implements that information to execute the functional behavior and purpose. By capturing original design intent, behavioral features ensure that product designs retain their integrity, robustness, and performance while adapting to market and engineering changes.

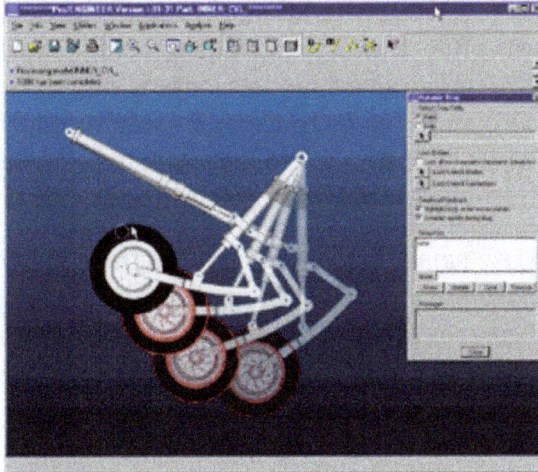

In one application, behavioral features can define the way that components in an assembly connect

## Flexible Evaluation

At the heart of Behavioral Modeling, adaptive process features make smart models highly flexible. As engineers make changes to smart models, the models regenerate to accommodate all their features and satisfy objectives and context. This regeneration may be fluid and dynamic, with a click-and-drag movement of a mechanism, for example. The regeneration may also be automated based on table-driven input; for instance, the process may select specific models from a generic family of parts or interchange alternate assemblies from an on-line parts catalog. Finally, the regeneration can occur as a result of program logic or driven by routines referenced in the feature itself. In all cases, this highly flexible adaptation allows smart models to respond to changes in their environment.

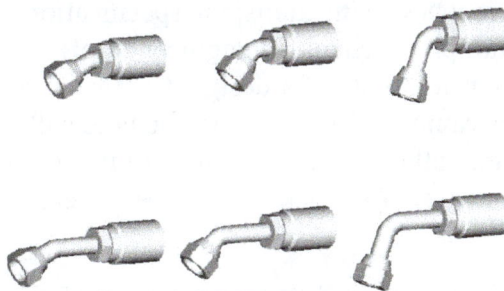

Change may be automatic and based on table-driven input

## Benefits of Smart Models

Because smart models encapsulate the key characteristics of a product, engineers no longer need to manage these aspects of the design; they are just part of the inherent intelligence of the model. For example, when application features such as joints and connections embody required constraints, they prohibit inconsistent behavior. Thus smart models allow engineers to focus on product differentiation and enhancement.

Since smart models facilitate design iteration, they promote design innovation. In the past, design iteration was often unachievable due to being too cumbersome, complex, or time consuming. For example, smart models make it easy to verify functional behaviors with click-and-drag operations or to check in-service interference after changes have been made. Such ease of iteration is critical, since many times innovative ideas are the result of investigating or validating something altogether different.

Organizations can also use smart models to increase their responsiveness to customer requirements. Using adaptive process features, engineers can create a springboard from which to perform custom product development. They can build intelligent products and then quickly and easily adapt those designs to new requirements.

## Objective-driven Design

An objective-driven design approach automatically optimizes designs to meet any number of objectives captured in the smart model by adaptive process features. It can simultaneously resolve conflicting objectives, a task that was often impossible using traditional approaches. Because the specifications are inherent in the features of the smart model, engineers can quickly and easily regenerate and revalidate the design's conformance to specifications as often as the model changes.

## Problem to be Solved

Traditionally engineers have maintained engineering and functional specifications in their heads, as notes on sketches, or in formal requirements documents. However, with Behavioral Modeling, specifications and requirements are captured and defined within the smart model. Rather than iterate through small changes to transform an initial design into a solution that meets specifications, the specifications can be used to actually drive the design. For example, perhaps an engineer wants to place a hole coincident with the axis of the center of gravity of a design. Capturing the center of gravity in a behavioral feature and parametrically tying it to the hole will ensure that the features remain coincident, even as other design changes are made and the center of gravity moves to reflect these changes. Moreover, this can be undertaken to meet any number of objectives.

In addition to defining the problem with standard types of measurements such as the center of gravity or an edge length, more complex requirements can be captured in behavioral features. For example, when designing the pattern of reflected light on a surface, not only is construction geometry required to accurately define the travel of the light from the source, to the reflector and onto the surface, but it is also needed to create surface normal and reflection angles. Other examples of nonstandard measurements could include mathematical equations that define the shrinkage of gas in a pipe as it cools. In Behavioral Modeling, these more complex requirements are captured in such a way that they can be shared among engineers in a library of features for reuse when needed.

## Understanding the Impacts of Change

The objective-driven design approach supports sensitivity studies, which give engineers multiple ways to assess and understand the impact of change on product design and performance. Sensitivity studies demonstrate how changes in design variables affect performance. For example, given a behavioral feature that specifies the volume of an automobile gas tank, a sensitivity study can illustrate the affect of modifications to the tank on its volume.

Design evaluations are easier when engineers can view results of their efforts in graphical displays that reside directly within the design environment. For instance, an engineer may want to click-and-drag a mechanism through its range of motion and visually detect potential interferences right on the screen. Or when studying the clearance between two loosely interfacing parts, an engineer may want to refer to an XY graph of position versus clearance to quickly understand the effect of a change in relative position.

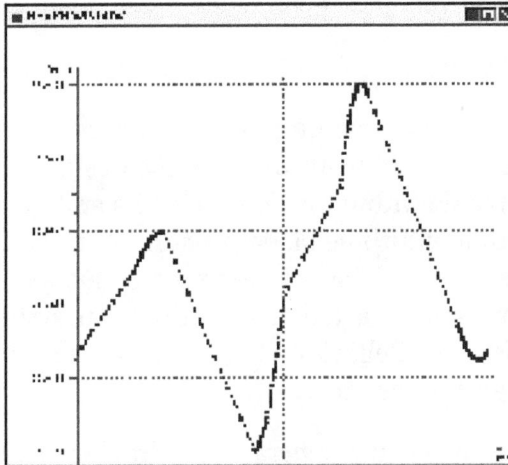

Graphs residing directly within the design environment communicate
the impacts of change to the engineer

## Synthesizing Multi-objective Designs

Objective-driven design identifies feasible and optimal solutions for complex problems by synthesizing multiple objectives. It accomplishes this goal by automatically searching the realm of all solutions to yield a set of feasible solutions. Through this automated design exploration, engineers can ascertain that a design meets all its engineering requirements. Moreover, engineers can use objective-driven design to find the single design that optimizes one or more goals, such as desired mass, minimum surface area, or maximum tolerance. Traditionally, engineers would need to manually try out as many of these iterations as they could within the time and resource constraints of the project.

## Benefits of Objective Driven Design

Because objective-driven design satisfies engineering specifications automatically, engineers can concentrate on designing higher performance, more functional products. Assured of a solution that meets basic design goals, engineers are free to use creativity and skill to improve the design.

## Open Extensible Environment

An open extensible environment is the third cornerstone of Behavioral Modeling. To maximize the benefits of the Behavioral Modeling approach, technology must be in place that allows engineers to take advantage of external systems, applications, information, and processes already in use within an organization. These external resources can contribute to the process of solving for design objectives and can return the results so that they become part of the final design. By providing connectivity throughout unique engineering processes, an open extensible environment increases design flexibility and results in more reliable designs.

## Connecting Applications

At the design level, smart models accommodate features that link to information in other applications. These external features make the design solution infinitely extensible. External features reside within the smart models and link to other applications, for example, the generation of turbine blade profiles by an external system during the modeling of a jet engine. External features permit engineers to regenerate and re-use designs built on live information created in many disparate systems. each time an engineer regenerates the model, the Behavioral Modeling system will reference and execute the external features as part of the process.

At the process level, this universal regeneration technology embeds closed-loop communication with external applications. It pushes attributes, parameters, and geometry from the model to the external routines and accepts the results as features native to the product model.

At the architecture level, a rich, concise, and easy-to-use protocol connects the applications seamlessly. This protocol also permits engineers to redefine proprietary programs so that they reside directly within a product model. For example, an engineer can re-implement a routine that aligns the center of mass to an axis of rotation into a model of a motor.

## Benefits of an Open Extensible Environment

The seamless engineering process afforded by an open extensible environment protects the product from loss of design intent. It avoids distractions brought about by cumbersome, disconnected communications and ensures the integrity of design information by directly connecting the 3D design definition to information that drives it.

# Boundary Representation

Boundary representation is one of the two most popular and widely used schemes to create solid models of physical objects. A B-rep model or boundary model is based on the topological notion that a physical object is bounded by a set of faces. These faces are regions or subsets of closed and orientable surfaces. A closed surface is one that is continuous without breaks. An orientable surface is one in which it is possible to distinguish two sides by using the direction of the surface normal to point to the inside or outside of the solid model under construction. Each face is bounded by edges and each edge is bounded by vertices. Thus, topologically, a boundary model of an object is comprised of faces, edges, and vertices of the object linked together in such a way as to ensure the topological consistency of the model.

## Topological and Geometrical Information of B-rep

The database of a boundary model contains both its topology and geometry. Topology is created by performing Euler operations and geometry is created by performing Euclidean calculations. Euler operations are used to create, manipulate, and edit the faces, edges, and vertices of a boundary model as the set (Boolean) operations create, manipulate, and edit primitives of CSG models. Euler operators, as Boolean operators, ensure the integrity (closeness, no dangling faces or edges, etc.) of boundary models. They offer a mechanism to check the validity of these models. Geometry includes coordinates of vertices, rigid motion and transformation (translation, rotation, etc.), and metric information such as distances, angles, areas, volumes, and inertia tensors. It should be noted that topology and geometry are interrelated and cannot be separated entirely. Both must be compatible otherwise nonsense objects may result. Figure3 shows a square which, after dividing its top edges by introducing a new vertex, is still valid topologically but produces a nonsense object depending on the geometry of the new vertex.

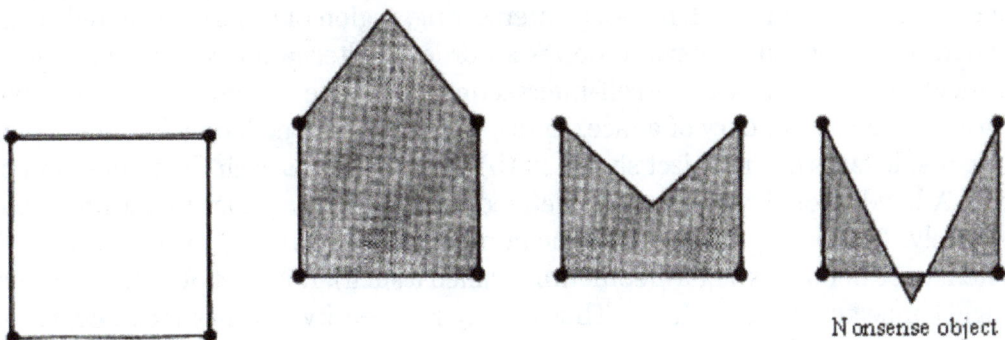

Nonsense object

(a) Original object                                  (b) Modified object

Effect of topology and geometry on boundary models

## Primitives of B-rep

If a solid modeling system is to be designed, the domain of its representation scheme (objects that can be modeled) must be defined, the basic elements (primitives) needed to cover such modeling domain must be identified, the proper operators that enable the system users to build complex objects by combining the primitives must be developed, and finally a suitable data structure must be designed to store all relevant data and information of the solid model.

Objects that are often encountered in engineering applications can be classified as either polyhedral or curved objects: A polyhedral object (plane-faced polyhedron) consists of planar faces (or sides) connected at straight (linear) edges which, in turn, are connected at vertices. A cube or a tetrahedron is an obvious example. A curved object (curved polyhedron) is similar to a polyhedral object but with curved faces and edges instead.

The reader might have jumped intuitively to the conclusion that the primitives of a B-rep scheme are faces, edges, and vertices. This is true if we can answer the following two questions. First, what is a face, edge, or a vertex? Second, knowing the answer to the first question, how can we know that when we combine these primitives we would create valid objects?

Polyhedral objects can be classified into four classes: The first class is the simple polyhedral. The second class is similar to the first with the exception that a face may be bounded by more than one loop of edges. The third class includes objects with holes that do not go through the entire object. The fourth and the last class includes objects that have holes that go through the entire objects. Topologically, these through holes are called handles.

With the above physical insight, let us define the primitives of a B-rep scheme. A vertex is a unique point (an ordered triplet) in space. An edge is a finite, non-self-intersecting, directed space curve bounded by two vertices that are not necessarily distinct. A face is defined as a finite connected, non-self-intersecting, region of a closed oriented surface bounded by one or more loops. A loop is an ordered alternating sequence of vertices and edges. A loop defines a non-self-intersecting, piecewise, closed space curve which, in turn, may be a boundary of a face. In (a), each face has one loop while the top and the right side faces of the object shown in (b) have two loops each (one inner and one outer). A handle (or through hole) is defined as a passageway that pierces the object completely. The topological name for the number of handles in an object is genus. The last item to be defined is a body (sometimes called a shell). It is a set of faces that bound a single connected closed volume. Thus a body is an entity that has faces, edges, and vertices. A minimum body is a point. Topologically this body has one face, one vertex, and no edges. The object on has two bodies (the exterior and interior cubes) and any other object in has only one body.

(a) Simple polyhedra

(b) Polyhedra with faces of inner loops

(c) Polyhedra with not through holes

(d) Polyhedra with handles(through holes)

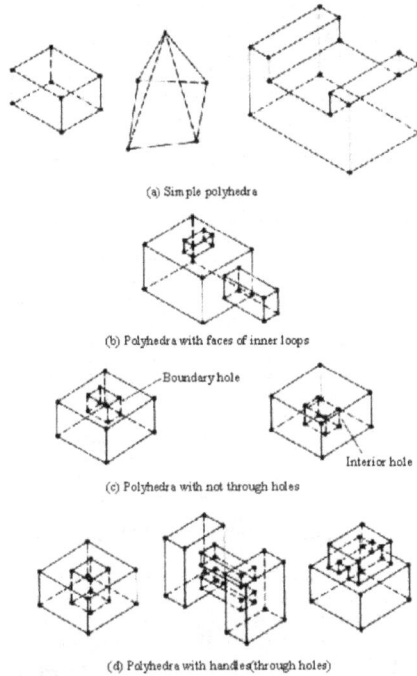

## Types of Polyhedral Objects

Faces of boundary models possess certain essential properties and characteristics that ensure the regularity of the model; that is, the model has an interior and a boundary. Faces are two-dimensional homogeneous regions so they have areas and no dangling edges. In addition, a face is a subset of some underlying closed oriented surface. At each point on the face, there is a surface normal N that has a sign associated with it to indicate whether it points into or away from the solid interior.

In traversing loops, the edges of the face outer loop is traversed, say, in a counterclock-wise direction and the edges of the inner loops are traversed in the opposite direction, say the clockwise direction.

## Euler's Law

Euler (in 1752) proved that polyhedra are topologically valid if they satisfy the following equation:

$$F - E + V - L = 2(B - G)$$

Where: F, E, V, L, B, and G are the number of faces, edges, vertices, faces' inner loop, bodies, and genus respectively. Eq. above is known as the Euler or Euler-Poincare law.

Open objects satisfy the following Euler's law:

$$F - E + V - L = B - G$$

(a) Wire polyhedra

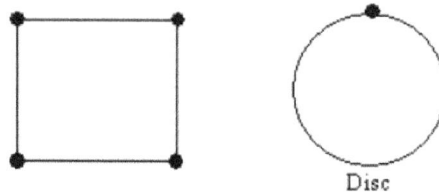

Disc

(b) Lamina polyhedra

Open polyhedral objects

## Exact B-rep and Faceted B-rep

The boundary model of a sphere has one face and one vertex. We now turn from polyhedral objects to curved objects such as cylinders and spheres. As shown in the figure below, the boundary model of a cylinder has three faces (top, bottom, and cylindrical face itself), two vertices, and three edges connecting the two vertices. The other "edges" are for visualization purposes. They are called silhouette edges.

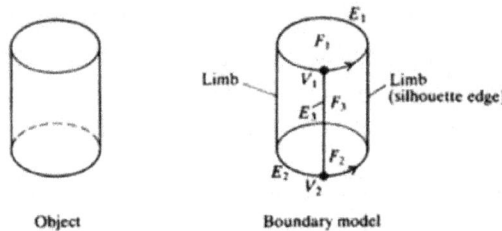

Object                 Boundary model

(a) Cylinder

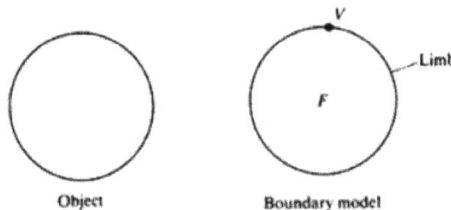

Object                 Boundary model

(b) Sphere

## Exact B-rep of a Cylinder and a Sphere

If the curved objects are represented by storing the equations of the underlying curves and surfaces of the object edges and faces respectively, the resulting boundary scheme is known as an exact B-rep scheme. Another alternative is the approximate or faceted B-rep. In this scheme, any curved face is divided into planar facets—hence the name faceted B-rep.

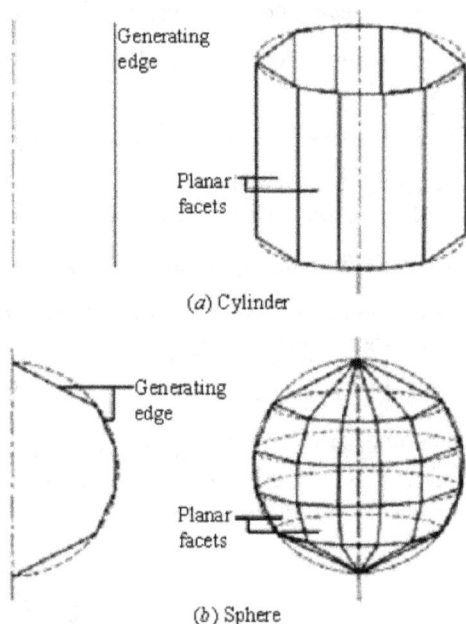

(a) Cylinder

(b) Sphere

Faceted B-rep of a cylinder and a sphere

## Advantages and Disadvantages of B-Rep

The B-rep scheme is very popular and has a strong history in computer graphics because it is closely related to traditional drafting. Its main advantage is that it is very appropriate to construct solid models of unusual shapes that are difficult to build using primitives.

Another major advantage is that it is relatively simple to convert a B-rep model into a wireframe model because the model's boundary definition is similar to the wireframe definition.

One of the major disadvantages of the boundary model is that it requires large amounts of storage because it stores the explicit definition of the model boundaries. It is also a verbose scheme—more verbose than CSG. The model is defined by its faces, edges, and vertices which tend to grow fairly fast for complex models. If B-rep systems do not have a CSG-compatible user interface, then it becomes slow and inconvenient to use Euler operators in a design and production environment.

## References

- Farin, Gerald (2002) Curves and Surfaces for CAGD: A Practical Guide, Morgan-Kaufmann, ISBN 1-55860-737-4

- Comprehensive-guide-product-design: smashingmagazine.com, Retrieved 31 March 2018

- Duggal, Vijay (2000). Cadd Primer: A General Guide to Computer Aided Design and Drafting-Cadd, CAD. Mailmax Pub. ISBN 978-0962916595

- Computer-aided-design-cad-and-computer-aided-cam: inc.com, Retrieved 15 March 2018

- Narayan, K. Lalit (2008). Computer Aided Design and Manufacturing. New Delhi: Prentice Hall of India. p. 4. ISBN 812033342X

- Understanding-caid-computer-aided-industrial-design: surfaceid.com, Retrieved 25 May 2018

- Hurst, J. (1989) Retrospectives II: The Early Years in Computer Graphics, pp. 39–73 in SIGGRAPH 89 Panel Proceedings, ACM New York, NY, USA, ISBN 0-89791-353-1doi:10.1145/77276.77280

- What-is-3d-modeling-2164: lifewire.com, Retrieved 16 May 2018

- "Looking Back: The TX-2 Computer and Sketchpad 82" (PDF). Lincoln Laboratory Journal. 19 (1). 2012. Archived from the original (PDF) on 2012-11-07

- 3d-modeling-things-youve-got-know: archicgi.com, Retrieved 26 June 2018

- What-is-architecture-rendering-194860901-20141130215742: linkedin.com, Retrieved 18 July 2018

# 4

# Production Support Machines

Production support machines are vital for industrial manufacturing processes for effective material handling, optimum production or for the execution of complex functions. Robots are vital production support machines. The topics elaborated in this chapter like industrial robots, articulated robots, SCARA robots, delta robots and Cartesian coordinate robots will help in providing a better perspective about such machines.

## Robots

A robot is a machine designed to execute one or more tasks automatically with speed and precision. There are as many different types of robots as there are tasks for them to perform.

Robots that resemble humans are known as androids, however, many robots aren't built on the human model. Industrial robots, for example, are often designed to perform repetitive tasks that aren't facilitated by a human-like construction. A robot can be remotely controlled by a human operator, sometimes from a great distance. A telechir is a complex robot that is remotely controlled by a human operator for a tele presence system, which gives that individual the sense of being on location in a remote, dangerous or alien environment and the ability to interact with it. Tele presence robots, which simulate the experience and some of the capabilities of being physically present, can enable remote business consultations, healthcare, home monitoring and childcare, among many other possibilities.

An autonomous robot acts as a stand-alone system, complete with its own computer (called the controller). The most advanced example is the smart robot, which has a built-in artificial intelligence (AI) system that can learn from its environment and its experience and build on its capabilities based on that knowledge.

Swarm robots, sometimes referred to as insect robots, work in fleets ranging in number from a few to thousands, with all fleet members under the supervision of a single controller. The term arises from the similarity of the system to a colony of insects, where the individuals and behaviors are simple but the fleet as a whole can be sophisticated.

Robots are sometimes grouped according to the time frame in which they were first widely used. First-generation robots date from the 1970s and consist of stationary, nonprogrammable, electromechanical devices without sensors. Second-generation robots were developed in the 1980s and can contain sensors and programmable controllers. Third-generation robots were developed between approximately 1990 and the present. These machines can be stationary or mobile, autonomous or insect type, with sophisticated programming, speech recognition and/or synthesis, and other advanced features. Fourth-generation robots are in the research-and-development phase, and include features such as artificial intelligence, self-replication, self-assembly, and nanoscale size (physical dimensions on the order of nanometers, or units of 10-9 meter).

## Essential Characteristics of a Robot

- Sensing First of all your robot would have to be able to sense its surroundings. It would do this in ways that are not un similar to the way that you sense your surroundings. Giving your robot sensors: light sensors (eyes), touch and pressure sensors (hands), chemical sensors (nose), hearing and sonar sensors (ears), and taste sensors (tongue) will give your robot awareness of its environment.

- Movement A robot needs to be able to move around its environment. Whether rolling on wheels, walking on legs or propelling by thrusters a robot needs to be able to move. To count as a robot either the whole robot moves, like the Sojourner or just parts of the robot moves, like the Canada Arm.

- Energy A robot needs to be able to power itself. A robot might be solar powered, electrically powered, battery powered. The way your robot gets its energy will depend on what your robot needs to do.

- Intelligence A robot needs some kind of "smarts." This is where programming enters the pictures. A programmer is the person who gives the robot its 'smarts.' The robot will have to have some way to receive the program so that it knows what it is to do.

## Reasons to use Robots

The reason robots are used is that it is often cheaper to use them over humans, easier for robots to do some jobs and sometimes the only possible way to accomplish some tasks! Robots can explore inside gas tanks, inside volcanoes, travel the surface of Mars or other places too dangerous for humans to go where extreme temperatures or contaminated environments exist. Robots can also do the same thing over and over again without getting bored. They can drill, they can weld, they can paint, they can handle hazardous materials, and in some situations, robots are much more accurate than a human - which can cut back on production costs, mistakes or hazards. Robots never get sick, don't need sleep, don't need food, don't need to take a day off, and best of all they don't ever complain! There are a lot of benefits to using robots.

## Parts of a Robot

Robots can be made from a variety of materials including metals and plastics. Most robots are composed of 3 main parts:

1. The Controller - also known as the "brain" which is run by a computer program. Often, the program is very detailed as it give commands for the moving parts of the robot to follow.

2. Mechanical parts - motors, pistons, grippers, wheels, and gears that make the robot move, grab, turn, and lift. These parts are usually powered by air, water, or electricity.

3. Sensors - to tell the robot about its surroundings. Sensors allow the robot to determine sizes, shapes, space between objects, direction, and other relations and properties of substances. Many robots can even identify the amount of pressure necessary to apply to grab an item without crushing it.

All of these parts work together to control how the robot operates.

# Industrial Robots

An industrial robot is an automatically controlled, reprogrammable, multipurpose manipulator programmable in three or more axes.

Because they can be programmed to perform dangerous, dirty and/or repetitive tasks with consistent precision and accuracy, industrial robots are increasingly used in a variety of industries and applications. They come in a wide range of models with the reach distance, payload capacity and the number of axes of travel (up to six) of their jointed arm being the most common distinguishing characteristics.

In both production and handling applications, a robot utilizes an end effector or end of arm tooling (EOAT) attachment to hold and manipulate either the tool performing the process, or the piece upon which a process is being performed.

The robot's actions are directed by a combination of programming software and controls. Their automated functionality allows them to operate around the clock and on weekends—as well as with hazardous materials and in challenging environments—freeing personnel to perform other tasks. Robotic technology also increases productivity and profitability while eliminating labor-intensive activities that might cause physical strain or potential injury to workers.

An industrial robot is made of three main components:

- Mechanics which represent what you actually see of the robot. This part is important for robot's precision, speed and payload.

- Electronics allows the command system to drive all motors and get environment information from sensors.

- Computer science which makes the robot "intelligent" through the collaboration with the robot environment and the user.

## Main Parts of an Industrial Robot

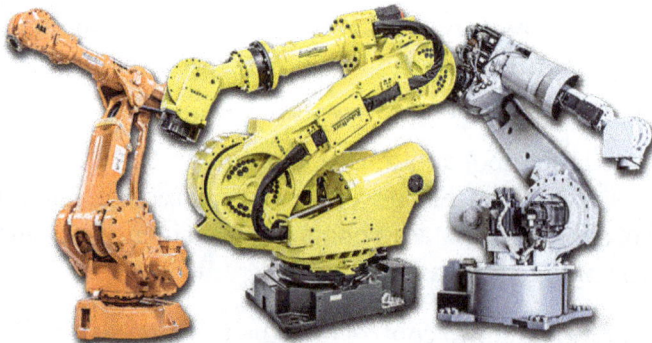

An industrial robot arm includes these main parts: Controller, Arm, End Effector, Drive, Sensor.

The controller is the "brain" of the industrial robotic arm and allows the parts of the robot to operate together. It works as a computer and allows the robot to also be connected to other systems. The robotic arm controller runs a set of instructions written in code called a program. The program is inputted with a teach pendant. Many of today›s industrial robot arms use an interface that resembles or is built on the Windows operating system.

Industrial robot arms can vary in size and shape. The industrial robot arm is the part that positions the end effector. With the robot arm, the shoulder, elbow, and wrist move and twist to position the end effector in the exact right spot. Each of these joints gives the robot another degree of freedom. A simple robot with three degrees of freedom can move in three ways: up & down, left & right, and forward & backward. Many industrial robots in factories today are six axis robots.

Classification of industrial robots by mechanical structure

The end effector connects to the robot›s arm and functions as a hand. This part comes in direct contact with the material the robot is manipulating. Some variations of an effector are a gripper, a vacuum pump, magnets, and welding torches. Some robots are capable of changing end effectors and can be programmed for different sets of tasks.

The drive is the engine or motor that moves the links into their designated positions. The links are the sections between the joints. Industrial robot arms generally use one of the following types of drives: hydraulic, electric, or pneumatic. Hydraulic drive systems give a robot great speed and strength. An electric system provides a robot with less speed and strength. Pneumatic drive systems are used for smaller robots that have fewer axes of movement. Drives should be periodically inspected for wear and replaced if necessary.

Sensors allow the industrial robotic arm to receive feedback about its environment. They can give the robot a limited sense of sight and sound. The sensor collects information and sends it electronically to the robot controlled. One use of these sensors is to keep two robots that work closely together from bumping into each other. Sensors can also assist end effectors by adjusting for part variances. Vision sensors allow a pick and place robot to differentiate between items to choose and items to ignore.

## Capabilities of Industrial Robots

Industrial robots are used in a variety of applications. These include:

- Handling: Capable of manipulating products as diverse as car doors to eggs, industrial robots are fast and powerful as well as dexterous and sensitive. Applications include pick and place from conveyor line to packaging, and machine tending, where raw materials are fed by the robot into processing equipment such as with injection molding machines, CNC mills and lathes and presses.

- Palletizing: Industrial robots load corrugated cartons or other packaged items onto a pallet in a defined pattern. Robotic palletizers rely on a fixed position or overhead gantry robot with special tooling that interfaces with the individual load components, building simple to complex layer patterns on top of a pallet that maximize the load's stability during transport. There are three primary types of palletizing: inline or layer forming, depalletizing or unloading, and mixed case.

- Cutting: Due to their dangerous nature, laser, plasma and water jet cutters are frequently used with robots. Hundreds of different cutting paths can be programmed into the robot, which produces precise accuracy and path following with greater flexibility than most dedicated cutting machines.

- Finishing: Multi-axis robots can grind, trim, fettle, polish and clean almost any part made in any material for a consistent quality finish.

- Sealing and gluing: To apply sealant or glue, a robot follows a path accurately with good control over speed while maintaining a consistent bead of the adhesive substrate. Robots are frequently used for sealing applications in the car industry to seal in windows, as well as in packaging processes for automated sealing of corrugated cases of product.

- Spraying: Due to the volatile and hazardous nature of solvent-based paints and coatings, robots are used in spray applications to minimize human contact. Paint robots typically have thin arms because they don't carry much weight, but need maximum access and movement fluidity to mimic a human's application technique.

- Welding: Used for both seam (MIG, TIG, arc and laser) and spot welding, robots produce precise welds, as well as control parameters such as power, wire feed and gas flow.

## Uses of Industrial Robots

Robots are used in a variety of ways throughout manufacturing and distribution.

- Load building: Assembling a pallet load of products at the end of a production line

- Manufacturing: Performing processing and assembly functions to work-in-process

- Quality control: Testing and inspection procedures deploy robots for repetitive or dangerous work

- Transportation: Loading pallets prior to shipping

- Warehousing: Removing received products from pallets and routing them to storage locations within a facility.

## Benefits of Industrial Robots

Industrial robots provide a variety of benefits:

- Accuracy – Robotic palletizers are software-directed for proper load placement

- Flexibility – Robotic systems can be re-purposed for other uses; end effectors can be switched out to handle different load types

- Lower labor costs – Automated pallet building reduces worker strain and frees operators for other tasks

- Quiet operation – Servo-based, robotic palletizers generate low noise levels

- Reduced product damage – Gentle handling prevents package and product damage

- Speed – The systems increase rate productivity up to 50%.

## Applications of Industrial Robots

Industrial robots are used in many industries, including:

- Aerospace
- Automotive
- Beverage
- Computers
- Consumer goods
- E-Commerce
- Electronics
- Food
- Grocery
- Hardware
- Healthcare
- Liquor distribution
- Manufacturing
- Medical products
- Pharmaceutical
- Quality control and inspection
- Retail
- Warehousing and distribution.

## Articulated Robots

Articulated Robots have all the joints of single axis revolute or rotary type. This chain of revolute joints provides greater freedom and dexterity in movement of the articulated robotic arm. SCARA and PUMA are the most popularly used articulated robots in assembly lines and packaging processes.

When people speak of industrial robots today, they are usually talking about articulated robots. These are the robots that are most commonly in use in factories worldwide. There are many different companies that produce articulated robots, including FANUC, Motoman, KUKA, Universal Robots, and ABB robotics companies.

Articulated robotic systems usually have four to six axes, but they can have up to 10. These robots have more degrees of freedom than any other robots on the market, which gives the manufacturer more versatility and makes them more appealing. An articulated robot can be used for dozens of different welding, material handling, dispensing and material removal applications.

So, why choose an articulated robot? Well, that is pretty simple. An articulated robot, like the FANUC R-2000iA, is a robot that can be used for several different applications, as mentioned above, and has a large work envelope. These robots are able to improve a company's productivity through improved speed and accuracy, which also improves the quality of the product being produced. Also, because of their popularity and their longevity, articulated robot systems are dropping in price, making them more affordable than ever before for other markets.

## Structure

Cylindrical Robot base has one revolute joint and two prismatic joints. Spherical Robot base has two revolute joints and one prismatic joint. After these two types, further the robot base is generalized as Articulated Robot base with all the joints being revolute joints. Any robot with revolute joints between all its base members falls in the category of articulated robot. The number of members in an articulated robot can be anything from two to ten. The axes of the revolute joints can be parallel or orthogonal to each other with some pairs of joints parallel and others orthogonal to each other.

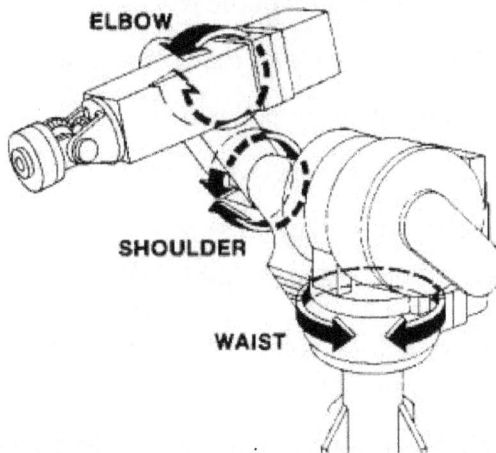

Articulated robots have a base called as waist which is vertical to the ground and the upper body of the robot base is connected to the waist through a revolute joint which rotates along the axis of the waist. Another link is also connected to the waist through a revolute joint perpendicular to the waist joint. This joint between the waist and the link is called as Shoulder of the articulated robot and the link may be called as the Arm.

One more link is connected to the arm through a revolute joint parallel to the shoulder joint. This joint with the arm forms the elbow of the articulated robot. Finally a wrist and a gripper is attached to the last link. The structure of the articulated robot with three revolute joints is very much similar to the human arm.

## 6-axis Robots

6-axis robots, or articulated robots, allow for articulated and interpolated movement to any point within the working envelope.

- Axis 1 - Rotates robot (at the base of the robot)

- Axis 2 - Forward / back extension of robot's lower arm

- Axis 3 - Raises / lowers robot's upper arm

- Axis 4 - Rotates robot's upper arm (wrist roll)

- Axis 5 - Raises / lowers wrist of robot's arm

- Axis 6 - Rotates wrist of the robot's arm.

Movement is driven by servo motors. A control system controls the power supplied to each motor and precise movement of the robot.

AdvantagesArticulated robots, or 6-axis robots, are easier to align to multiple planes, simple to operate and maintain, and easily redeployed for plastic injection molding automation applications on various types and sizes of plastic injection molding machines and for a wide range of upstream and downstream applications.

# SCARA Robots

Industrial robots are defined as 'multi-functional manipulators designed to move parts through various programmed motions'. As such, robots provides consistent reliable performance, repetitive accuracy and are able to handle heavy work loads and perform in harsh environments. Additionally, robots can be quickly reprogrammed to reflect changes in production needs and cycles.

The SCARA acronym stands for Selective Compliance Assembly Robot Arm or Selective Compliance Articulated Robot Arm.

Four axis SCARA robot that consists of an inner link that rotates about the World Z-axis, connected to an outer link that rotates about a Z elbow joint, which in turn is connected to a wrist axis that moves up and down and also rotates about Z. An alternative configuration has the linear Z motion as the second axis. These are the most popular geometries for vertical assembly and small parts pick-and-place operations.

SCARA are specifically designed for peg board type assembly and are heavily used in the electronics industry. They are very stiff in the vertical direction but have a degree of compliance in the horizontal plane that enables minor errors in the placement of components to be accounted for. These robots tend to be fairly small and capable of operating at high speed. They are used for assembly, palletisation and machine loading.

Scara robots have motions similar to that of a human arm. These machines comprise both a 'shoulder' and 'elbow' joint along with a 'wrist' axis and vertical motion. Scara's are ideal for a variety of general-purpose applications requiring fast, repeatable and articulate point to point movements such as palletizing, depalletizing, machine loading/unloading and assembly. Due to their 'elbow' motions, scara robots are also used for applications requiring constant acceleration through circular motions like dispensing and in-place gasket forming. Scara's are cylindrical robots, having two parallel rotary joints and provide compliance in one selected plane.

'Because of their speed, ruggedness, and durability, scara's generally are the first choice of manufacturers,' notes Peter Cavallo, robotics section manager at Denso Robotics of Long Beach, CA. 'Scara's are rigid, but less so than cartesians. All joints are at the end of the arm, resulting in more unsupported mass, which leads to more deflection.' The result, Cavallo says, is that scara robots are less able to cope with unsupported mass and their forces than cartesian systems.

The shoulder and elbow joints on scara's rotate around the vertical axes. The scara configuration provides substantial rigidity for the robot in the vertical direction, but flexibility in the horizontal plane. This makes it ideal for many assembly tasks, such as in the automotive and electronics industries, along with spray-painting and weld sealing applications.

Other applications of scara's can be found in the manufacture of fiber optics. 'In fiber optics, dust can be as large as the part itself. One of our customers had a field of view 0.7 mm by 0.5 mm and the robot was able to pick a part from this area,' says John Clark of Seiko/ Epson.

Scara robot joints are all capable of rotation and can be thoroughly sealed and safe-guarded which is necessary should the robot be deployed in dusty or corrosive environments, or for applications under water. 'Scara's are quicker, and can multiply motion at joints,' observed Stephen Harris, president of Rixan Associates, Inc. of Dayton, OH. Harris went on to say that scara's are used by their customers for electronic assembly. 'Scara's are also ideal for hip/knee replacement surgery. For this application, a modi-fied robot bores holes in the femur with a router bit. Usually a surgeon must bore this out by hand, then back-fill any extra,' remarked Steven Glor of Sankyo. The vice presi-dent of the Boca Raton, Fl-based Sankyo added, 'With this system, a three dimensional computer model of a patient's hip is made, and a robot uses this as a guide to bore out a precise hole in the bone. A scara is used because it is smaller and more easily transport-ed from one operating room to another. This cannot be done as easily with a cartesian.' Other surgical applications of scara's are to assemble radioactive isotope 'seeds' for treating prostate cancer. The seed is less than 6.5 mm in diameter and the assembly procedure is similar to putting lead into a mechanical pencil.

## Working of SCARA Robots

The SCARA robot is a manipulator with four degrees of freedom. This type of robot has been developed to improve the speed and repeatability on pick & place tasks from one location to another or to speed and improve the steps involved in assembly. This is why it is often used with FlexiBowl.

Morover these robots are used in the automotive field, as well as electronic, and other industrial fields where manufacturers needs to feed bulk components of all sizes; in these areas FlexiBowl is performing so good that is replacing others parts feeders.

SCARA kinematics makes this robot particularly suitable to perform assembly tasks with tight tolerances, such as putting a shaft into a hole, thanks to the capability to adjust the movement on the horizontal plane, while at the same time maintaining high rigidity on the vertical direction.

The mechanics of the SCARA arm is generally quite strong and can withstand without problems unexpected stress and collisions.

In a SCARA robot junctions of the shoulder and elbow are vertical and the wrist moves vertically. This configuration minimizes the effects of gravity on the robot by down-loading them to the ground and allowing the use of this machine in cases of strong pressures as in vertical perforations.

## Advantages

The SCARA robot is most commonly used for pick-and-place or assembly operations where high speed and high accuracy is required. Generally a SCARA robot can operate at higher speed and with optional cleanroom specification. In terms of repeatability, currently available SCARA robots can achieve tolerances lower than 10 microns, compared to 20 microns for a six-axis robot. By design, the SCARA robot suits applications with a smaller field of operation and where floor space is limited, the compact layout also making them more easily re-allocated in temporary or remote applications.

## Limitations

SCARA robots, due to their configuration are typically only capable of carrying a relatively light payload, typically up to 2 kg nominal (10 kg maximum). The envelope of a SCARA robot is typically circular, which doesn't suit all applications, and the robot has limited dexterity and flexibility compared to the full 3D capability of other types of robot. For example, following a 3D contour is something that will be more likely fall within the capabilities of a six-axis robot.

## Delta Robots

Biomimicry is the strategy of modeling designs and structures in technology after nature. Building a two-armed assembly robot to have wrist, elbow and shoulder joints, like a human, is a good example of biomimicry. In the design of robotic systems, engineers often look to the musculoskeletal structures of animals and humans to develop their designs. After all, if you are designing a robot to perform tasks usually done by human arms and hands, robotic arms and hands are a good starting point.

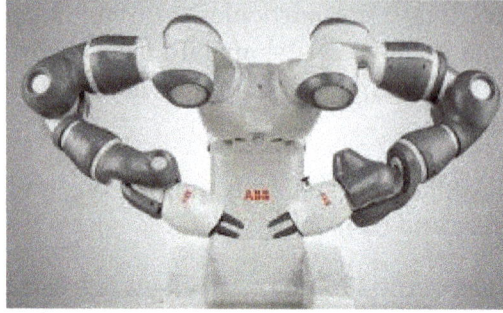

The design of this collaborative assembly robot is inspired by human physiology

What makes the delta robot fascinating is that it represents a complete departure from the constraints of biomimicry. It's an efficient, optimized machine, inspired not by nature, but by pure mathematics and geometry.

Today, delta robots are well established in the automation industry. Unlike larger articulated arms, delta robots are often kept as a stock item at many manufacturers such as Yaskawa Motoman and ABB. But in 1985, few robots could perform light pick-and-place tasks quickly or repeatably.

Reymond Clavel and his team at the Robotics Systems Laboratory at Ecole Polytechnique Federale de Lausanne (EPFL) began the research that would produce the delta robot following a visit to a chocolate factory. Clavel's team was looking for repetitive labor applications for robots, and they found that the packaging of chocolate pralines was a candidate for this type of high-speed, low-payload automation.

Clavel's team began by setting constraints on their design. First, the robot must perform at a rate of 3 picks per second. In order to place the chocolates correctly, the mobility of the robot required 4 degrees of freedom: translations along 3 axes, as well as rotation about the vertical axis. In order to achieve a high rate of work, Clavel added two more constraints to the design: the actuators of the robot would be fixed on the frame, and the moving part of the robot would be kept as light as possible.

The first prototype of the delta robot

Six months after the visit to the chocolate factory, a prototype of the delta robot was complete; by December, a patent was filed. Two years later, the delta robot was industrialized by a small company called Demaurex Robotics and Microtechnology.

While engineers and industry players were impressed with the innovative design, first reactions to the new robot were tepid. They would not take the first step and risk their reputation with a robot similar to an umbrella." As with any new technology, potential customers demanded to know how the product would facilitate their operations. What's the ROI?

Thirty-two years later, industrial professionals are no longer dubious of these umbrella-shaped robots, which have become unparalleled in pick-and-place, sorting, and other high-speed, low-mass applications. However, they're still asking about how delta robots can benefit their production process, and for the hard numbers of the ROI.

## Delta Robot Configurations

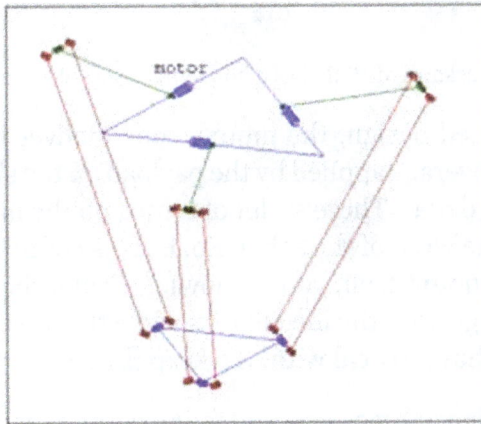

Delta robot kinematics

Here's how a typical delta robot works: Three high-torque servomotors are mounted on a rigid frame. On each motor shaft, an arm is mounted perpendicular to the shaft's rotational axis. Some robots use direct drive, in which the arm is mounted directly on the shaft, and some use a gearbox. These "Bicep" arms are connected to lightweight linking rods arranged in parallelograms to restrict twisting motion. These arms are connected to a central platform. The joints at both ends of each parallel rod move freely, typically in ball joints. At the lower platform, the end effector may be attached, as well as several other options including motors for additional axes of freedom. Most delta robots have, at least, a fourth axis, enabling objects to be rotated.

The main benefit of delta robots is that the heavy motors are fixed on the frame, allowing the moving parts of the robot to be very light. In contrast, each motor of an articulated arm robot carries the weight of all the successive motors. In the mechatronic design of delta robots, the motion is being translated down through, in most cases, carbon fiber arms, where there's far less mass being moved. They've become a very, very efficient way of x, y, and limited z translation."

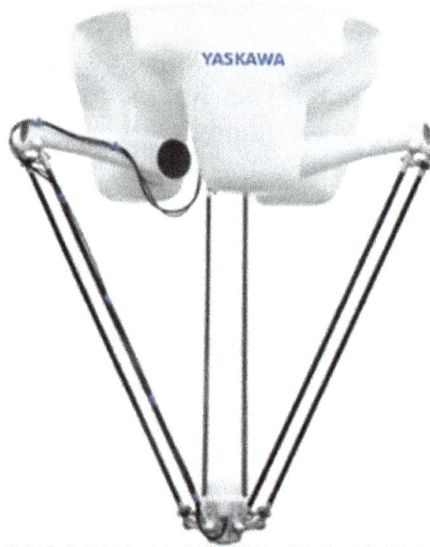

The Yaskawa MPP3H has a horizontal reach of 1300mm

Because of this lever-based design, the motors must deliver high torque in order to counteract the increase leverage applied by the payload on the shaft. This contributes to the low payloads of most deltas. There's a lot of torque on the motors that gets translated all the way down to the wrist of the robot. So, because of that offset, because of that distance away from the motor itself, you're moving a wand that can be 1,300 millimeters or even longer in length. So, the idea here is while the motors are high-torque, you have this offset that you have to deal with. To keep the speed, you limit the payload."

While most delta robots have three arms and four axes of freedom, there are other configurations available. Omron, for example, offers the Quattro parallel robot. According to the company, this design allows the robot to carry an increased payload, or to perform a faster cycle time.

This brings up an interesting point about theta motion around the z axis, which is the fourth axis of freedom for the delta robot. There are two main design approaches to this. The traditional solution would be to have three motors mounted above, and then a separate motor to do rotation. Omron has that. You can mount the motor on top or at the platform. But if you mount the motor on the top, then you must have a drive shaft. You can mount the motor at the platform, but it will be a smaller motor and the added weight will increase the inertia of the platform."

The shafted configuration allows certain benefits, including higher payload. "We have a number of patented designs for the theta shaft and the theta fully rotational shaft, which can move in continuous 360-degree rotation. The reason we perform theta motion by a shaft, and not by adding a motor at the central plate, is that such a motor adds weight and reduces payload and reliability. The shafted design avoids these issues." The reliability issue is with the cable that must run to the motor. The cable must not interfere with the high-speed arms and must stand up to repetitive bending and twisting strain.

On the other hand, the shafted approach adds another element which requires maintenance. The shaft can wear out over time, and needs regular replacement, and in some cases, regular maintenance such as greasing.

## Axes a Delta can Manipulate in

The range of motion of delta robots is simpler than that of a serial manipulator robot. Delta robots can move freely within their workspace in the x and y directions. Z-axis motion is more limited. However, depending on the application, rotations about the x, y and z axes are sometimes required. "It's not uncommon anymore to see rotational axes or pitching axes to upright parts, for example. If you were to look at a blow molding application where you perhaps wanted to descramble horizontally oriented bottles and upright them into a soldiering or singulation type of arrangement, you can pitch those parts by adding external axes and then you could actually add in many cases, another roll axis."

The FANUC M-1iA/0.5A can manipulate objects in six axes

"The majority of operations taking place with delta robots are going to require a four axis. There are certainly applications that require that six-axis because a six axis will let you, for example, flip a small bottle up or something similar. You could pick something laying down and place it standing straight up. That's where you see that kind of robot."

Payload can be another indicator of your best robot option. "In theory, the delta model could be designed to handle a higher payload, but the heavier design is going to start reducing the cycles per minute. At that point it may not be worth all the complex mechanics of a delta robot. From a reliability or price standpoint, a SCARA robot could be more effective at a higher payload. For example, some SCARA robots can handle 20kg.

## Number of Robots one Need

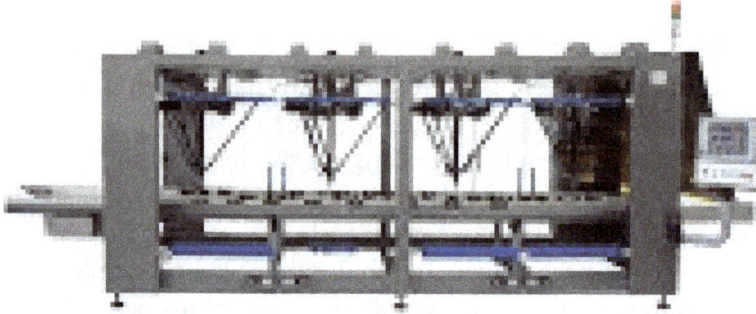

This can be a tricky question for fledgling automation customers to answer. Consider the following:

- Form factor and work envelope

- Vision system requirements

- Programming and controls

Beyond these secondary factors, the robot you need is essentially a function of your needed cycle rate and the mass of your products. This information is available on robot datasheets and from integrators.

Delta robots excel at providing high-speed pick-and-place, but they are sadly still bound by the laws of physics, especially Newton's first law. Some delta robots can produce accelerations of up to 12-15 g. "Because delta robots start and stop so fast, when you decelerate with the product in hand, if you don't have enough grasping power, the product can be released and go flying. In some applications you need very high power vacuum etc. There are certain products you cannot grip using suction. You have to use mechanical grippers. Of course, mechanical grippers add weight to the end effector, so the capacity is reduced of Kawasaki Robotics.

End effectors are a key part of selecting the best solution for the job: "When you calculate the cycle time, you have to look at the payload that you want to move, and the cycle that you need to achieve. You don't want to move any faster than you absolutely need to. The key point is designing the gripper effectively to hold onto your parts under your given acceleration. The design of the gripper is always the key to that."

While the most common end effectors for pick-and-place robots are suction cup grippers due to their fast actuation and high repeatability, not all objects can be manipulated by suction. For example, objects with porous or rough surfaces are typically manipulated by a mechanical gripper. However, mechanical grippers reduce cycle time, as the fingers must actuate open and closed to pick and release each object.

## Reasons to use Delta Robots

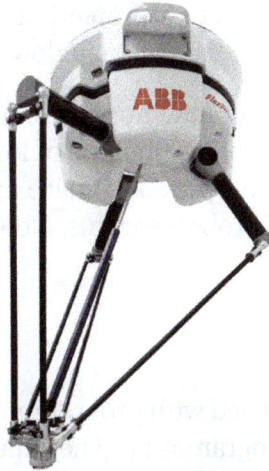

The ABB IRB 360 FlexPicker

"They're certainly becoming a mainstream alternative for high speed picking and placing applications. It was that for the longest time we were limited to slower articulated arm technology or slightly faster SCARA technology, and now with Delta robots we're able to get the heavy components up in the air, onto a super structure, and really increase speed because in many cases we're moving much less mass," "So, they're great for when you've got to go fast,". "This is really about looking at the architecture of the robot. Due to global megatrends, production lines are getting faster. Because of the parallel kinematic architecture, the delta robot has a very high mechanical advantage in performance and flexibility when compared to traditional articulated robots, and even Cartesian robots".

Delta robots are purpose-built for pick-and-place, going back all the way to their inception with Dr. Clavel. "Typically you'd need a Delta robot if you have rate that's going to be sustained—a lot of parts flowing down the conveyor belt, hundreds of parts per minute. There are other robots that can do fast rates, but they can't keep up with the rate of a delta without overheating or working over their motor duty."

Delta robots are on the rise in the industrial pick-and-place market. "I see a growing use of Delta robots, especially in packaging and in high speed pick-and-place. I think in the future, we will see them moving into different industries for lightweight objects that require high speed pick-and-place. So, I see them spreading more," Massoud predicted.

## Delta Robots in Industrial Food Production and Packaging

Of course, there are a few applications beyond pick-and-place for delta robots. Many of these applications are in the food industry, an industry characterized by high volume and strict health and safety regulations. For example, food manufacturers must meet standards for equipment wash down, and prevent exposure of food product to potential contaminants such as lubricants, fragments of metal or plastic, or dust. Because the architecture of the delta robot, the motors can be well isolated in enclosures, and many models are available up to IP69K ingress protection rating, allowing the robot to be blasted by high-pressure, high-temperature wash down.

Here's a fascinating example of a unique application: "One customer has robots in big bakeries to do bread scoring, scoring the tops of loaves of bread as they go by on a conveyor. That's one example of non-pick-and-place application,". "Traditionally, one of our integrators has used six-axis robots for this. Now, the same integrator is looking at using our delta robot."

## Delta Robot Programming

In general, any technician experienced with programming industrial robots will be fully equipped to handle delta robot programming. The experts agreed that delta robots are in some ways simpler to program than serial manipulators, and in some ways, more complex.

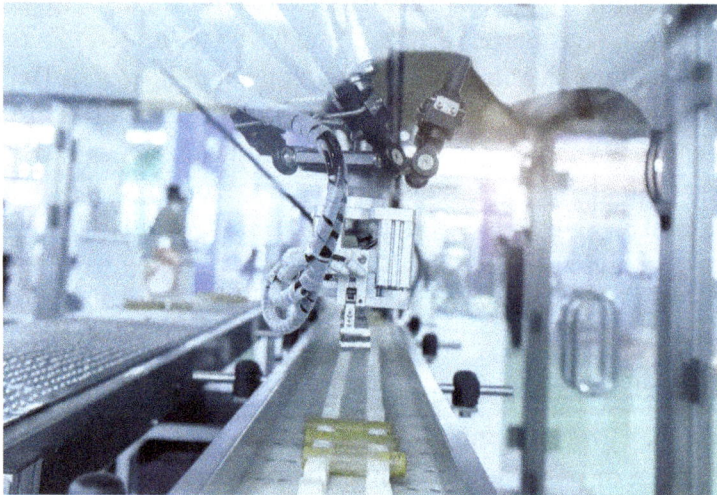

Working at high speeds, delta robots are carefully programmed to work in synchronization with the conveyors parts move along. The conveyor may be equipped with an encoder to pass information to the PLC. In order to organize efficient picking of these parts, a vision system is used to give the correct parts to the robots, which is also controlled by the PLC.

Where it gets really interesting is in cases where multiple robots work together to pick from the same line. Software uses the vision data to schedule picks among the robots to optimize the duty of each robot.

Besides these complexities, the robots are, in the end, programmed according to their spatial coordinates, in x, y, and z. Most manufacturers use the same language and even the same environment to program and control delta robots as they use for all their industrial robots. FANUC, for example, uses the same interface for both four axis delta robots and their six axis arm robots. Zanotti noted that if you try to enter a value that requires movement in the fifth or sixth axes, the software will simply return a fault.

Here's an excellent description of the role of vision in the function of these robots:

"The purpose of the vision system is to identify good from bad products. We use a variety of algorithms to filter out the good products and give those to the robots. We have a range of cameras on offer, including 3D cameras. 3D vision is exciting because it can open up particular benefits for data analysis of products. There was a time where we were only using grey scale cameras, now we're predominantly color, and in the future, we will be moving even further into 3D cameras."

Today's vision systems can identify objects of different shapes, sizes and even colors, enabling sorting and quality operations.

## Common Failure Modes and Maintenance

One of the advantages of these robots is that the design is open and easy to access for repair and maintenance. We asked each expert about the most common wear and tear issues to watch out for.

Most of the experts agreed that due to the repetitive motion, the ball and socket joints connecting the parallel arms were subject to wear and eventual replacement. Luckily, it's an easy fix.

"Most of these robots have a ball and socket construction, where the carbon fiber arms either meet the main drive at the top of the robot, or the wrist at the bottom of the robot. Inside of those drives there's typically an insulator ring that's made of a nylon or Delrin, and that provides a friction bearing type of arrangement inside that joint. Those usually will get changed out about every six months," "Other than that, it's your traditional greasing of some Zerk fittings on the motors,". "Assuming that a cell's running normally, the wear is going to be in the linkages in the parallel links coming down from the base of the robot. Follow a recommended maintenance plan to check the certain wear points. Washers, springs, that sort of thing,". "If they are worn, they're usually cheap, non-invasive fixes that don't take too long. They don't bring your line down for too long.", those rings typically last up to one million cycles before needing replacement.

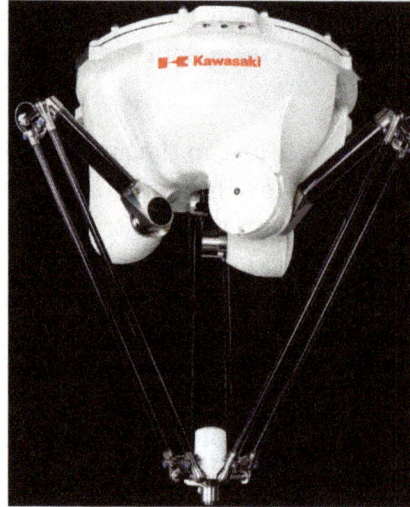
Kawasaki YF003N Delta

"In general, the main motors wear at the same rate. It depends on the motion pattern that repeats, but in general the three motors work within a reasonable range. AC servo-motors driving the arms are one of the most reliable motor products for very fast start stop motion. So, there are no brushes, you have a magnetic rotor rotating inside of your stator windings. There are hardly any components inside to wear,"

## Safety

Safety for delta robots does not differ from requirements for other industrial robots. "You definitely need guarding around the delta robot. It could be Plexiglas, it could be wire mesh structure. At the same time, on long large lines, where you can't get anywhere close to the delta robot because of the conveyor in feed/exits, many have seen open layouts with nothing around the delta, because you can't reach into the envelope."

One thing to consider is that with product being accelerated to high speeds, it's possible that a suction or mechanical gripper could fail during motion, and send the payload flying. Therefore, it's important to ensure guarding protects against this risk.

# Cartesian Coordinate Robots

With automation on the rise, the case for linear robotics, has expanded. Linear robots are a type of industrial robot with two or three principal axes that move in a straight line rather than rotate, functioning at right angles to each-other. The three sliding joints correspond to moving the wrist; up and down, back and forth, as well as in and out. Linear robots with horizontal members supported at both ends are referred to as Gantry robots.

Because there are no rotating axes, linear robotics tend to have a higher degree of accuracy, making them the ideal automation solution for those mundane repetitive tasks. Unlike other automation machines, linear robotics systems can be reprogrammed to accommodate product changes quickly, and are flexible to meet unique requirements. A linear robot can be more economical than other types of robots such as articulated arm or SCARA.

Many industries today require automated processes to ensure process repeatability, reduce variability, meet quality expectations and lower manufacturing costs. Automation is vital to many companies to meet lead-time demands and remain competitive. Robotic systems offer the best cost investment as they can be reprogrammed or repurposed to meet product variety and life cycles. Linear robotics are a versatile solutions for these challenges.

There are many different uses for Linear robotics, but the most common are:

- Pick and Place Solutions – High-speed pick and place robots move a product from one location to another with spot-on accuracy. Manual errors like placing the wrong item in the wrong place can be replaced with linear robotics. Implementing a pick and place system to place items in certain spots and on assembly lines can greatly improve efficiency and accuracy, as well as prevent injuries.

- Sorting – A linear robot can streamline the traditionally manual and monotonous process of sorting, making it more efficient and safer. When coupled to a vision system, accurate distinctions can be made with greater consistency.

- Packaging Solutions – Packaging processes can run 24/7 when utilizing a linear robotics system because you can theoretically run even with the lights out! Allowing production to run non-stop would be extremely difficult for laborers to manually handle. This eliminates the need for a 3rd shift crew. This is a great way to improve turnaround time.

- Palletizing – A palletizer takes products and places them in a predetermined pattern to form layers and then place them onto a pallet. Without automation, this could be a dangerous heavy-lifting manual job. Robotic systems greatly increase productivity and reliability for palletizing processes. They have a minimal equipment footprint and have become a more ideal solution for a wider range of packaging scenarios.

- Assembly Processes – Many processes such as dispensing, cutting, forming, welding etc. are better performed by a Linear Robotic system, especially when long travels and extended reach are required.

## Cylindrical Base Robot

Robots mounted on base bodies with revolute, cylindrical or spherical joints along with prismatic joints have better functionality and move with ease between the points in space.

The body of this type is such that the robotic arm can move up and down along a vertical member. The arm can rotate about that vertical axis and the arm can also extend or contract. This construction makes the manipulator able to work in a cylindrical space. The dimensions of the cylindrical space are defined as, radius by the extent of the arm and height by the movement along the vertical member. The cylindrical manipulator base body has one revolute joint at the fixed frame, one cylindrical joint about the axis of rotation and one prismatic joint in the arm of the manipulator.

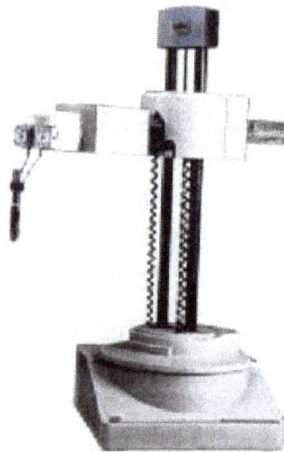

The position of the end is defined by the extension of the arm, height of the arm and rotation of the main body axis. These are the three variables to be controlled to position the end effectors of a cylindrical base robot. In other words this type of structure forms a cylindrical coordinate system and be controlled the same way.

## Workspace of Cylindrical Base Robot

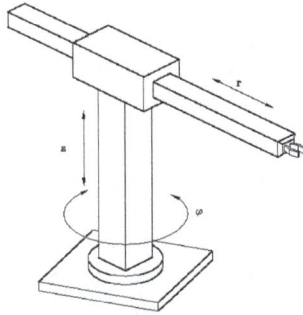

The reach of the end of the Cylindrical Robot is a cylinder whose dimensions are determined by the motion limits of the different parts of the robot. But the motion limits of any joint in on the both sides, maximum as well as the minimum.

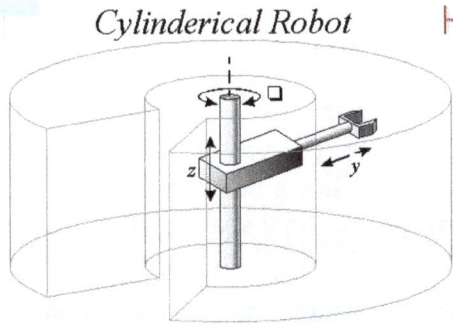

Cylinderical Robot

Thus, the workspace, volume comprised of the points where the end point of the robotic arm can be positioned, is not a complete cylinder but it is an intersection of two concentric cylinders. Dimensions of the inner cylinder are determined by the minimum limits of the motion of robot parts.

## Replacing a Cartesian Robot with a Cylindrical Robot

Both the Cartesian and Cylindrical base robots, being able to reach points in the three dimensional space, can be interchanged having common minimum workspace for the application. Each robot base has its own suitable applications. For some applications Cartesian may be more preferred and for other applications Cylindrical Base Robots may be suitable. Even then, the two types can be interchanged with some advantages or disadvantages.

To count as an advantage, Cylindrical Base Robots can move between the required points faster than the Cartesian Robot, especially in the case when the two points are at the same radius. Out of the three motions two are along the same axis. And for disadvantage one can consider the efforts required to make transformation of instructions from the Cartesian coordinate system to the cylindrical coordinate system.

# Spherical Base Robot

Spherical Base of the robots has just one replacement of the joint in the Cylindrical Base of the robots. The slider joint between the robot arm and the vertical member is replaced by a revolute joint. This change of a joint makes the spherical based robot more dexterous than cylindrical robot.

Spherical bases of the robots make them capable of working in a spherical space. Though the workspace cannot be more than a three dimensional one but with the increasing number of the revolute joints the arm movements of the robot can become more sophisticated. Spherical Bases Robots, as the name says, work in a space defined by the intersection of spherical volumes. With wide range of possibilities of complex movements robots with spherical bases find application in many industrial processes.

## Construction of the Base Structure

Spherical type of manipulator has the base member which can rotate about the vertical axis. A member is connected to the base member through a revolute joint and this member can extend and contract like a telescope. This arrangement of the base body makes the manipulator arm to work in a space defined as the intersection of spherical spaces.

The spherical base has the same, three, numbers of joints as the other three dimensional robot bases has. Two joints are revolute joints and the remaining is a prismatic joint such that the arm of the robot can extend and retract. The end effectors of the robot are mounted on this telescopic arm. The two revolute joint movements can be actuated by direct coupling with the servo motors and the telescopic arm movement can be actuated by a rack and pinion arrangement. Spherical base has three degrees of freedom and three variables to be controlled to operate it.

## Reach and Workspace of Spherical Base Robot

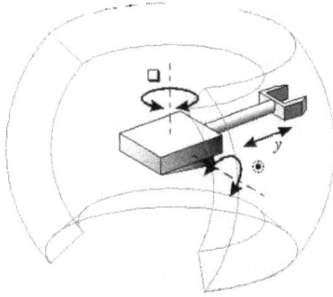

Reach of a robot is the limits to which its end effectors can approach the work points. For the spherical robots the reach of its end effectors is determined by the limits of the motion of the three joints of the base. For such type of base of the robots the reach of the end effectors of the robot is a sphere. The radius of this sphere is dependent on the maximum extension of the telescopic arm.

The workspace of the spherical base robots is the volume intersection of the two concentric spheres. The dimensions of the external sphere are equal to the maximum limits of the joint movements and the radius of the inner sphere is determined by the minimum limits of the joint movements which are in turn governed by design constraints. The range of rotation of the revolute joint at base and between the base member and the arm determines the sector of the sphere that can be covered; and the range of movement of the telescope arm determines the range of the radius of the spherical volume of intersection.

Spherical Robots are more involved in construction and more dexterous in working so is there control. The control of spherical robots requires three variables as Cartesian and Cylindrical robots do but the coordinate frame and there transformation is bit complex than other types.

Spherical Robots can perform tasks requiring movement in three dimensional spaces easily. It can reach at points in the space with ease, speed and accuracy. But to make such three dimensional robot work properly according to the specified instruction of reaching particular points and doing job there it requires a substantial amount of background work involving calculations of coordinate frames and control variables.

## Work Frame of Spherical Robots

It is helpful to define coordinate frames at suitable points on the structure of a robot to analysis and effect the movements of different parts and the robot as a whole. The primary frame, common to any type of robot, is the coordinate frame fixed to the base of the robot. This frame is called as the world coordinate frame or can be called as global coordinate frame. The points in the workspace are provided in this frame. In spherical robots the second frame is attached to the joint between the vertical member and the arm. Third frame is attached to the wrist of the robot.

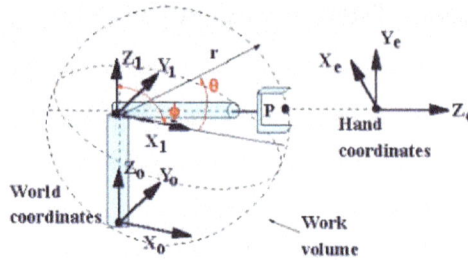

## Control Coordinates of Spherical Robots

The target points for the end effectors are specified as per the task in world coordinate frame. But to make the end effectors move to the specific points the actuators attached to each joints have to be provided with the values of their respective movements producing the final effect as desired. This means that the control variables are the values provided to the actuators, hence, the coordinates of the target points have to be converted to the control coordinates.

The global coordinates and the control coordinates are represented in the matrix form. Control Coordinates are obtained from Global Coordinates through various transformations of the matrices. For a particular transformation of a matrix it is multiplied with a transformation matrix to obtain the required transformation. All these calculations are incorporated in a computer program and implemented through micro controllers and processors such that it takes the global coordinates and supply the actuators with the required values of the control coordinates.

## Applications of Spherical Base Robots

The type of robot base to choose for a particular application is primarily decided by the reach and workspace requirement of the task. Then we search for a robot base with similar reach and workspace. After this selection criterion other details such as the accuracy and repeatability of the robot are considered. The common applications of spherical base robots are in material handling operations in the production line, such as, transferring or pick-and-place materials and stacking objects.

## Industrial Robots for Different Applications

The application is the type of work that the robot is designed to do. Robot models are created with specific applications or processes in mind. Different applications will have different requirements. For instance, a painting robot will require a small payload but a large movement range and be explosion proof. On the other hand, an assembly robot will have a small workspace but will be very precise and fast. Depending on the target application, the industrial robot will have a specific type of movement, linkage dimension, control law, software and accessory packages. Below are some types of applications.

# Welding Robots

# Material Handling Robots

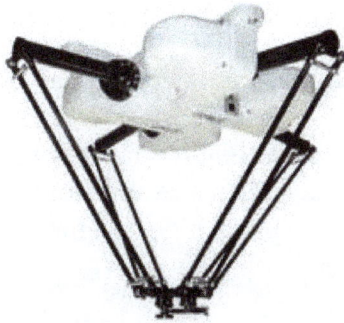

# Palletizing Robot

# Painting Robot

## Assembly Robot

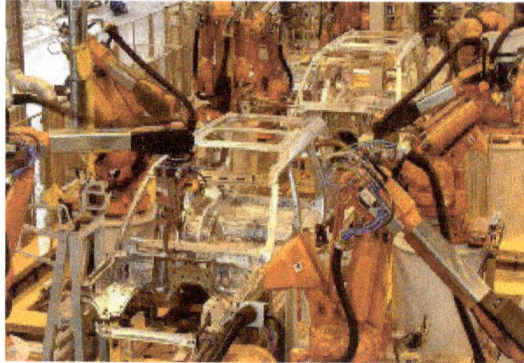

## Choosing the Best Industrial Robot for a Process

A simpler, more complete definition of robotic types can be narrowed down to five types: Cartesian, Cylindrical, SCARA, 6-Axis and Delta. Each industrial robot type has specific elements that make them best-suited for different applications. The main differentiators among them are their speed, size and workspace. Knowledge of each operating aspect of all five types can help machine designers choose the best robot for their process.

## Cartesian

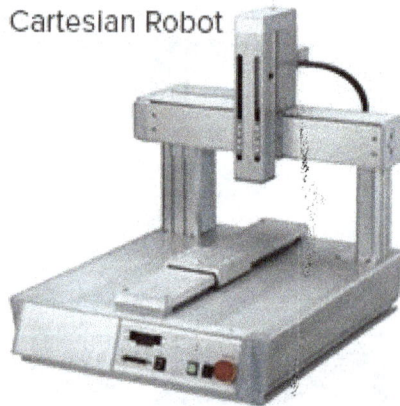

Cartesian Robot

The most commonly used robot type for the majority of industrial applications is Cartesian. Plant operators often default to this type because they are easy to use and program. The linear movements of the Cartesian elements give the robot a cube-shaped workspace that fits best with pick and place applications and can range from 100 milimeters to tens of meters. These robots are also a popular choice because they are highly customizable. Customers can determine the stroke lengths, speed and precision of the robots because most of the parts arrive separately and are assembled by the machine builders. That being said, one drawback to Cartesian robots is the complexity of assembly required. Overall, plant operators choose this robot design most often for the flexibility in their configuration that allows them to meet specific application needs.

## Cylindrical

Cylindrical robots are very simple and similar to Cartesian in their axis of motion. Most Cylindrical robots are made of two moving elements: rotary and linear actuators. Because they have a cylindrical work envelope, machine designers might select them for their economy of space. The robot can be placed in the middle of a workspace and, because of its rotation element, it can work anywhere around it. Simple applications where materials are picked up, rotated and then placed work best for Cylindrical robots. Installation and use are not complex, and they come as fairly complete solutions with minimal assembly.

## SCARA

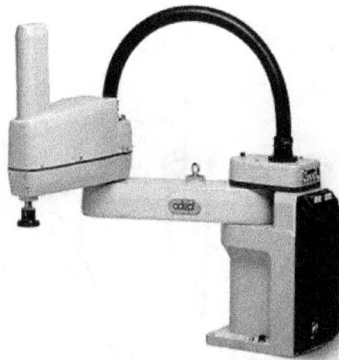

SCARA robots offer a more complete solution than the Cartesian or Cylindrical. They are all-in-one robots, meaning a SCARA robot is equipped with x, y, z and rotary motion in one package that comes ready-to-go, apart from the end-of-arm tooling. The work envelope is similar to Cylindrical robots but it has more degrees of motion in a radius or arch-shaped space. Applications are also similar to Cylindrical and Cartesian robots, but SCARA robots can move quicker than the other two. They are seen often in biomed applications due to their small work area. Because SCARAs have the easiest integration they seem like the best solution for the majority of applications, but Cartesians are more common because of their level of customization.

## 6-Axis

Another all-in-one robot type is the 6-Axis. Though sometimes 6-Axis robots can be almost toy-sized, they are typically very large and used for large assembly jobs such as putting seats into a car on an assembly line. These robots operate like a human arm and can pick up materials and move them from one plane to another. An example of this would be picking a part up from a table top and putting it into a cupboard— something the other robot types cannot do easily. 6-Axis robots can move quick and come in complete solutions like SCARAs, however, their programming is more complicated. The robots can get so large and move so quickly that, if roller coaster seats were attached to them, they could simulate an amusement park ride. Because they are one of the largest of the five robot types, most designers choose them for their ability to make movements that others cannot to compensate for the loss of space.

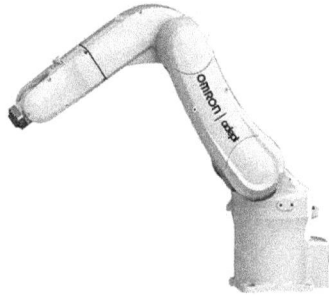

## Delta

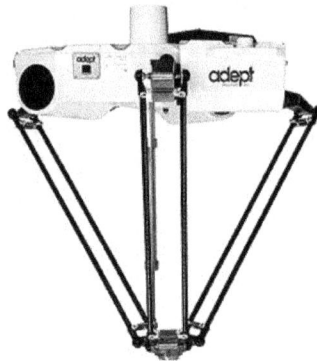

As the fifth and final type, Delta robots are the fastest and most expensive. They have a unique, dome-shaped work envelope in which they can achieve very high speeds. Delta robots are best for fast pick-and-place or product transfer applications like moving parts from a conveyor belt and placing them in boxes or onto another conveyor belt. They also come as complete solutions for machine designers but are more complicated in use than the 6-Axis or SCARA robots. The main advantage of Delta robots is their speed and precision with which they operate.

## Safety and Maintenance of Industrial Robots

Across the board, all five types of robots come with almost the same level of safety implications. The typical method of protecting an operator from getting pinched or hit is an external system that basically creates a fence around the robots. This fence is a hard guard with a gate that, when opened by an operator, tells the robot to stop moving or switch to a mode of slow operation. This hard guard protects both the operator and the product by not allowing anyone to tamper with the robot when it is in use. As far as maintenance goes, there is no standard across the board for the robotic types. Maintenance periods mostly depend on the environments in which the robots are operating and their duty cycle. For example, processes with heavy exposure to dirt or dust will require more frequent maintenance on all robot types than processes in clean rooms.

## Choosing the Best Fit

When designers are making the decision to implement one of the five robot types in their processes, they need to consider the basic starting points for any motion application: load, orientation, speed, travel, precision, environment and duty cycle. Determining these factors will draw direct correlations to the corresponding robot type that will give them the most efficient and effective results in their plant.

## References

- Articulated-robots-more-axes-equal-more-possibilities: robots.com, Retrieved 17 July 2018

- Base-bodies-of-robots-articulated-robot-base-29981: brighthubengineering.com, Retrieved 25 March 2018

- When-to-use-a-scara-robot, laser-systems: cyan-tec.com, Retrieved 16 June 2018

- What-are-linear-robotics, job-stories: macrondynamics.com, Retrieved 22 July 2018

- Base-bodies-of-robots-cylindrical-base-robot-29395: brighthubengineering.com, Retrieved 29 May 2018

- Spherical-base-robot-construction-and-workspace-29493: brighthubengineering.com, Retrieved 09 May 2018

- Base-bodies-of-robots-spherical-base-robot-control-and-application-29636: brighthubengineering.com, Retrieved 29 April 2018

- What-are-the-different-types-of-industrial-robots-63528: robotiq.com, Retrieved 25 April 2018

- Five-types-of-industrial-robots-and-how-to-choose-the-best-fit: valin.com, Retrieved 25 April 2018

# 5

# Computer Aided Production

The rapid advancement in electronic and automation technologies has brought tremendous innovation in computer aided production. This emerging branch of engineering studies such innovations in design and manufacturing. All the diverse topics central to the understanding of computer aided production such as computer numerical control machine, cutter location, direct numerical control, pencil milling, etc. have been exten-sively discussed in this chapter.

## Computer Aided Production Engineering

In the early 1990s, due to the economic recession, more industrial firms worldwide suffered a severe downturn in business. In the mid-'90s, with a renewed and booming, customer-driven market, business success is contingent upon coming to market faster with new products and building up output to meet the increasing demand. The automotive industry was one of the first industries to respond quickly to consumers diverse tastes and products shorter life cycles. On example the average development cycle of a vehicle was reduced from seven to five years, with three years as the current goal. But this goal cannot be achieved by compromising on quality: in todays competitive market, quality is no longer a selling feature but a basic element of any product. Thus, manufacturers have begun to look closely at their current methods and examine how they can increase their products offering while meeting three major challenges: shorter time to market, increased quality and lower manufacturing costs.

- Shorter time to market:

Shorter time to market requires more efficient and productive development tools. The same work needs to be done, not only quicker, but also with fewer errors. Design have to be "first time right" and efforts should be made towards catching errors up front and eliminating rework. Moreover, manufacturers can no longer afford to walk consecutively through the development phases, to make significant progress, they must make large strides towards concurrent engineering. These changes involve greater development risk, and tools to minimize these risks must be readily available. Standardizing the manufacturing tools will also reduce time to market, as it will eliminate lengthy redesigns, speed up production ramp-up time and reduce downtime through better maintenance.

- Increased quality:

Increased quality requires verification and analysis of the manufacturing processes to check that they comply with design intent. In addition, an optimized design of the manufacturing process delivers savings in process cycle time. This time may be used to enhance the manufacturing process in order to increase the products quality.

- Cost savings:

Cost savings can be generated through productivity gains, reduced capital investment, better allocation of manpower, efficient management of design changes and reduced overheads.

To decide where to invest to gain competitive edge, we have to study the impact of each investment on the complete development and manufacturing process. Over the past few years, manufacture invested heavily in the two ends of the industrial process:

- The product design phase by install CAD systems
- The production phase by application of automated devices.

But as these automated tools become industrial standards, manufacturers have to look elsewhere to maintain improvement and competitive edge. This is the reason why many top manufacturers are increasing their use of Computer Aided Production Engineering (CAPE) tools as part of Computer Integrated Manufacturing (CIM).

Computer Integrated Manufacturing (CIM) represent the integration of traditional production and engineering technologies with the computer technology, which enable the automation all activities from product design to their expedition (design of products, creation of technological procedures, production planning, operative control, manufacturing of products, quality control, assembly, packaging, expedition, etc.), with goal to bring down of the material and energy pretension, to increase of work productivity, to bring down of supplies, to shorten of development and production time, to increase of time and power utilize of production devices and ti increase of products quality.

The complex computer integration is not only goal or in many firms it is reality. bring down of computer components prices and increase of computer power in unite with modern software technologies, new methods of firm organisation, new progressive technologies condition orientation on modern information and communication technologies in many firms. The CIM systems in most cases is not represented by complex wholes, or they are compile by integration of partial automatized systems - CA systems (Computer Aided Systems), composition which is shown in figure below.

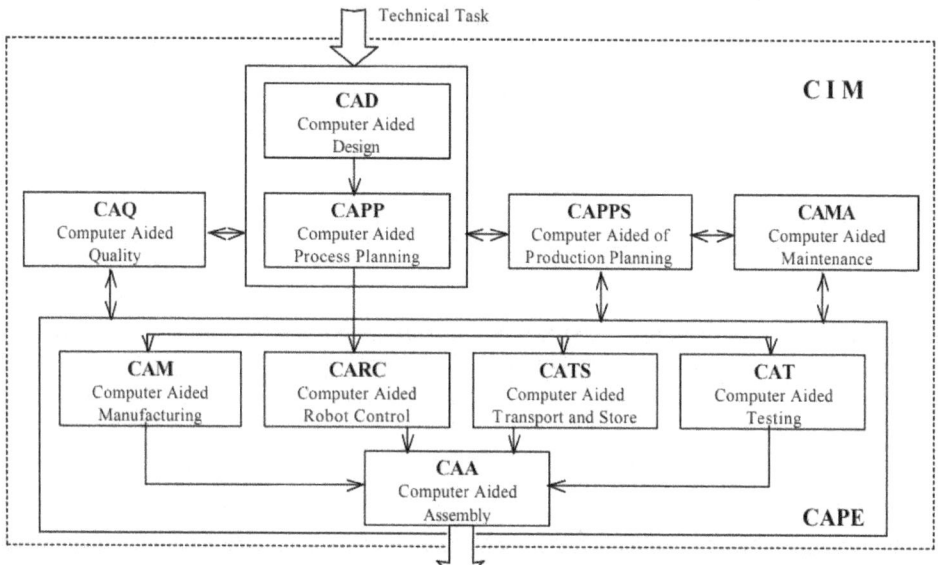

Organization of partially CA systems in CIM complex

The complex of CIM can be integrated by many partial CA systems, where familiar and more utilized are :

- CAD - Computer Aided Design,
- CAPP - Computer Aided Process Planning,
- CAQ - Computer Aided Quality,
- CAPPS - Computer Aided of Production Planning System,
- CAMA - Computer Aided Maintenance,
- CAPE - Computer Aided Production Engineering,
- CAM - Computer Aided Manufacturing on NC and CNC machine tools and devices for progressive technologies (laser, water jet etc.),
- CARC - Computer Aided Robot Control,
- CATS - Computer Aided Transport and Store,
- CAT - Computer Aided Testing,
- CAA - Computer Aided Assembly.

Computer Aided Production Engineering (CAPE) is a subsystem of the system CIM including the computer aided systems of all activities connected with realization of product manufacturing (programming of machine tools, manipulation, transport and store devices, measuring, testing and diagnose of parts and assembled product). This stage of computer aided systems in complex CIM fluently establish on application of computer aided systems in technical (construction and technological) preparing of production and is inevitable for secure of concurrent engineering conditions.

All of the above considerations lead to the conclusion that graphic computerized process planning holds huge potential for improvements on all fronts. This is even more so considering the fact that changing a product design is almost always less costly than changing the manufacturing process.

The enabling technologies for CAPE (Computer Aided Production Engineering) emerged only in the mid-'80s. Simulation, advanced graphics, motion emulation and powerful computers to support them all matured to the extent that CAPE technology could be brought into economically justifiable use.

## Present Time and Future of CAPE Technologies

CAPE started as an off-line programming tool for automated manufacturing equipment. Its prime purpose was:

- To program robots off the shop floor,
- To provide the operators a safer working environment,
- To an efficient tool to perform trial-and-error routines,
- A reduction in maintenance and troubleshooting efforts.

Better use of the production equipment for real manufacturing purposes rather than preparation work. Soon the benefits of using CAPE tools upstream became clear. Why use CAPE only for programming equipment, why not use it up-front, for designing the whole work cell? Enhanced CAPE tools enabled manufacturing engineers to design the complete work cell in a faster, optimized and error-free fashion. The ability to view the equipment working in a manufacturing environment allowed for much tighter designs with less error margins, as well as more accurate time and flow calculations.

Thus, CAPE took a significant step forward. Although savings resulting from off-line programming were significant, they were only in the initial phase of production. Computerized process design provided benefits not only in the launch phase, but throughout the product life cycle, as optimized cell layouts and tools paths resulted in reduced capital investment and lower variable manufacturing costs.

The example of computerized optimization of cutting tool path by simulation with aim of CAM system is on.

A typical task of a production engineer is to design a complete manufacturing process. Using a CAPE tool, he creates a graphical representation of a factory work cell in his computer. Then he imports the products geometric 3-D CAD data. He selects the appropriate production tools from electronic reference libraries, where all the capabilities and features of these tools are kept together with their respective geometric data. In this virtual manufacturing environment, he designs the process.

The drive for concurrent engineering has given birth to some new terms in the industrial world: digital mock-up, master model, electronic prototype - all names for the same concept - refer to a single database that contains all the product data, and is consistently updated with the latest changes. This database will allow all departments involved in the same project to work on the most recent data.

Example of module CIMULATOR utilize in CAD/CAM system Cimarron IT

At the front end of the industrial process, CAD integration technology allows CAPE to create the production master model by directly accessing the CAD master model. This assures that the production engineer's work on the most recent product data and that they can progress with their production design alongside the product designers. It actually facilitates a CAPE database that both complements the CAD database and holds all production-related data. At the back end of the development process, CAPE has to provide a smooth transition to the manufacturing equipment. With this aim, a European forum was established consisting of automated equipment vendors, CAPE and computer-aided robotics vendors, and their customers. The outcome is the RRS (Realistic Robot Simulation) standard interface which allows a more accurate and realistic emulation of the equipment. The example of automatized production work cell with two industrial robots activity viewing by simulation and programming system ROANS is on.

Based on installation of CAPE technologies we can production firms divide to two groups:

- Firms where has a longer time installed CAPE technologies as important tools of production engineering in industrial processes,

- Firms where managers are rapidly closing the gap, migrating to CAPE as they did to CAD in the 80s.

Realistic simulation of robotized work cell function by system ROANS

CAPE technologies has evolved from a simulation and offline programming tool to a mainstream production engineering tool which is tightly integrated with the other computerized and automated tools in the industrial development cycle.

# Computer Numerical Control Machine

Computer numerical control (CNC) is a method for automating control of machine tools through the use of software embedded in a microcomputer attached to the tool. It is commonly used in manufacturing for machining metal and plastic parts.

With CNC, each object to be manufactured gets a custom computer program, usually written in an international standard language called G-code, stored in and executed by the machine control unit (MCU), a microcomputer attached to the machine. The program contains the instructions and parameters the machine tool will follow, such as the feed rate of materials and the positioning and speed of the tool's components. Mills, lathes, routers, grinders and lasers are common machine tools whose operations can be automated with CNC. It can also be used to control non-machine tools, such as welding, electronic assembly and filament-winding machines.

CNC machines are electro-mechanical devices that manipulate machine shop tools using computer programming inputs.

CNC machining is a manufacturing process in which pre-programmed computer software dictates the movement of factory tools and machinery. The process can be used to control a range of complex machinery, from grinders and lathes to mills and routers. With CNC machining, three-dimensional cutting tasks can be accomplished in a single set of prompts.

Short for "computer numerical control," the CNC process runs in contrast to— and thereby supersedes — the limitations of manual control, where live operators are needed to prompt and guide the commands of machining tools via levers, buttons and wheels. To the onlooker, a CNC system might resemble a regular set of computer components, but the software programs and consoles employed in CNC machining distinguish it from all other forms of computation.

**CNC machining** is a manufacturing process in which pre-programmed computer software dictates the movement of factory tools and machinery.

When a CNC system is activated, the desired cuts are programmed into the software and dictated to corresponding tools and machinery, which carry out the dimensional tasks as specified, much like a robot.

In CNC programming, the code generator within the numerical system will often assume mechanisms are flawless, despite the possibility of errors, which is greater whenever a CNC machine is directed to cut in more than one direction simultaneously. The placement of a tool in a numerical control system is outlined by a series of inputs known as the part program.

With a numerical control machine, programs are inputted via punch cards. By contrast, the programs for CNC machines are fed to computers though small keyboards. CNC programming is retained in a computer's memory. The code itself is written and edited by programmers. Therefore, CNC systems offer far more expansive computational capacity. Best of all, CNC systems are by no means static, since newer prompts can be added to pre-existing programs through revised code.

## CNC Machine Programming

In CNC, machines are operated via numerical control, wherein a software program is designated to control an object. The language behind CNC machining is alternately referred to as G-code, and it's written to control the various behaviors of a corresponding machine, such as the speed, feed rate and coordination.

Basically, CNC machining makes it possible to pre-program the speed and position of machine tool functions and run them via software in repetitive, predictable cycles, all with little involvement from human operators. Due to these capabilities, the process has been adopted across all corners of the manufacturing sector and is especially vital in the areas of metal and plastic production.

For starters, a 2D or 3D CAD drawing is conceived, which is then translated to computer code for the CNC system to execute. After the program is inputted, the operator gives it a trial run to ensure no mistakes are present in the coding.

## Open/Closed-loop Machining Systems

Position control is determined through an open-loop or closed-loop system. With the former, the signaling runs in a single direction between the controller and motor. With a closed-loop system, the controller is capable of receiving feedback, which makes error correction possible. Thus, a closed-loop system can rectify irregularities in velocity and position.

In CNC machining, movement is usually directed across X and Y axes. The tool, in turn, is positioned and guided via stepper or servo motors, which replicate exact movements as determined by the G-code. If the force and speed are minimal, the process can be run via open-loop control. For everything else, closed-loop control is necessary to ensure the speed, consistency and accuracy required for industrial applications, such as metalwork.

In CNC machining, movement is usually **directed across X and Y axes.**

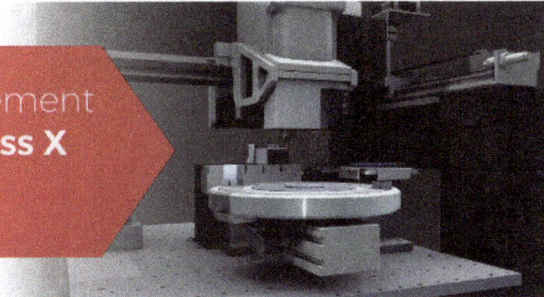

## CNC Machining as Fully Automated

In today's CNC protocols, the production of parts via pre-programmed software is mostly automated. The dimensions for a given part are set into place with computer-aided design (CAD) software and then converted into an actual finished product with computer-aided manufacturing (CAM) software.

Any given work piece could necessitate a variety of machine tools, such as drills and cutters. In order to accommodate these needs, many of today's machines combine several different functions into one cell. Alternately, an installation might consist of several machines and a set of robotic hands that transfer parts from one application to another, but with everything controlled by the same program. Regardless of the setup, the CNC process allows for consistency in parts production that would be difficult, if not impossible, to replicate manually.

## Types of CNC Machines

The earliest numerical control machines date to the 1940s when motors were first employed to control the movement of pre-existing tools. As technologies advanced, the mechanisms were enhanced with analog computers, and ultimately with digital computers, which led to the rise of CNC machining.

The vast majority of today's CNC arsenals are completely electronic. Some of the more common CNC-operated processes include ultrasonic welding, hole-punching and laser cutting. The most frequently used machines in CNC systems include the following.

## CNC Mills

CNC mills are capable of running on programs comprised of number- and letter-based prompts, which guide pieces across various distances. The programming employed for a mill machine could be based on either G-code or some unique language developed by a manufacturing team. Basic mills consist of a three-axis system (X, Y and Z), though most newer mills can accommodate three additional axes.

> Some of the more common CNC-operated processes **include ultrasonic welding, hole-punching and laser cutting.**

## Lathes

In lathe machines, pieces are cut in a circular direction with index able tools. With CNC technology, the cuts employed by lathes are carried out with precision and high velocity. CNC lathes are used to produce complex designs that wouldn't be possible on manually run versions of the machine. Overall, the control functions of CNC-run mills and lathes are similar. As with the former, lathes can be directed by G-code or unique proprietary code. However, most CNC lathes consist of two axes — X and Z.

## Plasma Cutters

In a plasma cutter, material is cut with a plasma torch. The process is foremost applied to metal materials but can also be employed on other surfaces. In order to produce the speed and heat necessary to cut metal, plasma is generated through a combination of compressed-air gas and electrical arcs.

## Electric Discharge Machines

Electric-discharge machining (EDM)— alternately referred to as die sinking and spark machining — is a process that molds work pieces into particular shapes with electrical sparks. With EDM, current discharges occur between two electrodes, and this removes sections of a given work piece.

When the space between the electrodes becomes smaller, the electric field becomes more intense and thus stronger than the dielectric. This makes it possible for a current to pass between the two electrodes. Consequently, portions of a work piece are removed by each electrode. Subtypes of EDM include:

- Wire EDM, whereby spark erosion is used to remove portions from an electronically conductive material.

- Sinker EDM, where an electrode and work piece are soaked in dielectric fluid for the purpose of piece formation.

In a process known as flushing, debris from each finished work piece is carried away by a liquid dielectric, which appears once the current between the two electrodes has stopped and is meant to eliminate any further electric charges.

## Water Jet Cutters

In CNC machining, water jets are tools that cut hard materials, such as granite and metal, with high-pressure applications of water. In some cases, the water is mixed with sand or some other strong substance. Factory machine parts are often shaped through this process.

Water jets are employed as a cooler alternative for materials that are unable to bear the heat-intensive processes of other CNC machines. As such, water jets are used in a range of sectors, such as the aerospace and mining industries, where the process is powerful for the purposes of carving and cutting, among other functions.

**CNC lathes** are used to produce complex designs that wouldn't be possible on manually run versions of the machine.

## Further Uses of CNC Machines

As plenty of CNC machine video demonstrations have shown, the system is used to make highly detailed cuts out of metal pieces for industrial hardware products. In additional to the aforementioned machines, further tools and components used within CNC systems include:

- Embroidery machines

- Wood routers

- Turret punchers

- Wire-bending machines

- Foam cutters

- Laser cutters

- Cylindrical grinders

- 3D printers

- Glass cutters

When complicated cuts need to be made at various levels and angles on a work piece, it can all be performed within minutes on a CNC machine. As long as the machine is programmed with the right code, the machine functions will carry out the steps as dictated by the software. Providing everything is coded according to design, a product of detail and technological value should emerge once the process has finished.

# CNC Machine Tool Monitoring by AE Sensors

## Acoustic Emission in Machining

In many manufacturing processes machining forces in tools and materials generate high-frequency acoustic emission (AE). Various investigations in recent years have shown that AE signals can be used to analyse and monitor machining operations. The use of acoustic emission techniques for process monitoring has the potential of ensuring high product quality while minimizing the total cost of a product. Processes such as cutting, grinding, forming and joining all generate acoustic emission for reasons unique to each process. In many cases, the emission can be monitored to characterize the process, to detect defects or process abnormalities in situ, and to detect finishing quality.

Principal areas of interest with respect to AE signal generation in metal cutting are in the primary generation zone ahead of the tool where the initial shearing occurs during chip formation. The secondary deformation zone is along the chip/tool rake face interface where sliding and bulk deformation occur. The third zone is along the tool flank face/work piece surface interface. Finally, there is a fourth area of interest, that associated with the fracture of chips during the formation of discontinuous chips.

In milling (rotating multiple toothed cutters) the cutting process is discontinuous with varying chip load and relative rubbing generated noise. Thus the basic AE signal due to the material deformation inherent in chip deformation and chip/tool contact is complicated by noise and periodic interruptions of the cutting process. This is in contrast to the relatively stationary single point turning generated AE.

The major sources of acoustic emission identified in metal cutting include:

- Plastic deformation of work piece material in the shear zone;
- Plastic deformation and sliding friction between chip and tool rake surface;
- Sliding friction between work piece and tool flank surface;
- Collision, entangling and breakage of chips.

In milling, because of the interrupted cutting conditions, additional sources include:

- Shock wave generated in tool entry;
- Sudden unloading and chip break off at tool exit.

In both cases tool velocity and metal removal rates are significant parameters affecting the energy of the AE signal. Because of the geometry of tool/work piece interaction, there are additional variations in the AE generated, mainly resulting from chip thickness and tool velocity changes as the tool rotates through the work piece material.

Acoustic emission is usually one of two distinct types: continuous (arriving at the transducer in such large numbers that distinct events cannot be detected) and burst signals (distinct emission events can be observed). In the machining process, for example, continuous acoustic emission signals are generated in the shear zone, at the chip/tool interface and at the tool flank/work piece surface interface while burst signals are generated by chip breakage during or after formation or by insert fracture.

## Acoustic Emission Sensors

The fundamental of a successful process monitoring system is the right selection of sensors. A wide variety of sensors have been utilized to monitor machining process. Most of the attentions are focused on force measurements and tool condition monitoring (TCM) which includes tool identification, tool wear monitoring, tool breakage and tool life. Most practical approaches to tool condition monitoring have been developed utilizing indirect measurements of tool performance that are easier to achieve than direct measurements in most environments.

Acoustic emission sensors are used on a test object's surface to detect dynamic motion resulting from acoustic emission events and to convert the detected motion into a voltage-time signal. This voltage signal is used for all subsequent steps in the acoustic emission technique. The electrical signal is strongly influenced by characteristics of the sensor and since the test results obtained from signal processing depend so strongly on the electrical signal, the type of sensor and its characteristics are important to the success and repeatability of acoustic emission testing.

A wide range of basic transduction mechanisms, capacitive transducers, displacement sensors or even the laser interferometers, can be used to detect acoustic emission. But acoustic emission detection is commonly performed with sensors that use piezoelectric elements for transduction. The element is usually a special ceramic such as lead zircon ate titivate (PZT), as shown in figure below and is acoustically coupled to the surface of the test item so that the dynamic surface motion propagates into the piezoelectric element. The dynamic strain in the element produces a voltage-time signal as the sensor output.

Surface motion of a point on a test piece may be the result of acoustic emission. Such motion contains a component normal to the surface and two orthogonal component tangential to the surface. Acoustic emission sensors can be designed to respond principally to any component of motion but virtually all commercial acoustic emission sensors are designed to respond to the component normal to the surface of the structure. Since shear and Rayleigh wave speeds typically have a component of motion normal to the surface, acoustic emission sensors can often detect the various waves.

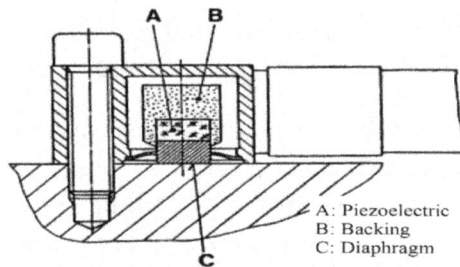

A: Piezoelectric
B: Backing
C: Diaphragm

## A Typical AE Sensor

Mostly acoustic emission sensing is based on the processing of signals with frequency contents in the range from 30kHz to about 1MHz. In some special applications, detection of acoustic emission at frequencies below 20 kHz or near audio frequencies can improve testing and conventional microphones or accelerometers are sometimes used.

Acoustic emission sensing often requires a couplant between sensor and test material. The purpose of the couplant is to provide a good acoustic path from the test material to the sensor. A sensor hold-down force of several Newton is normally used to ensure good contact and to minimize the couplant thickness.

Sensors for process monitoring must meet the following requirements:

- Measurement point as close to the machining point as possible
- No reduction in the static and dynamic stiffness of the machine tool
- No restriction of working space and cutting parameters
- Wear-free and maintenance-free, easy to change, low costs
- Resistant to coolant, dirt, chips and mechanical, electromagnetic and thermal influences

- Function independent of tool or work piece

- Adequate metrological characteristics

- Reliable signal transmission.

Figure below shows the experimental setup in this study. Both force/torque sensor and AE sensor were deployed in a PC controlled 3-axis milling machine. This report only discusses AE sensor and signal processing. A four-flute end mill cutter was chosen to machine an aluminum alloy material.

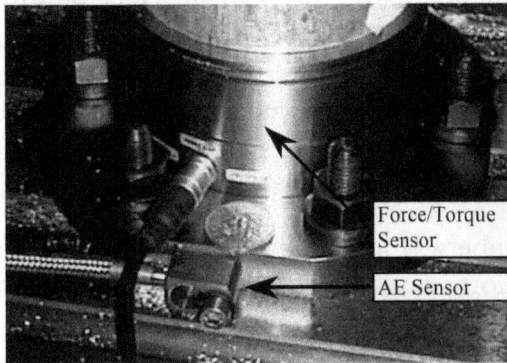

Force/torque and AE sensor mounted on the machine bed

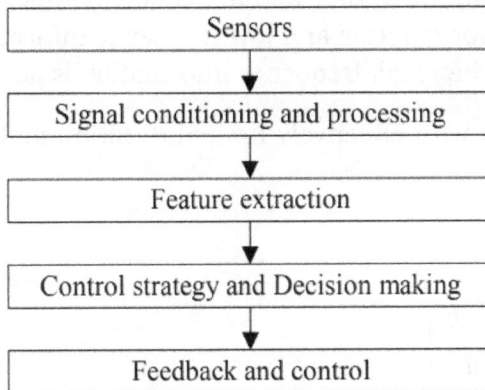

Components of monitoring and control system

Figure above shows all the components, which are required for monitoring and control of a machine tool. Perhaps the most important element of the control system is the signal processing methodologies for feature extraction and decision making.

## Advanced Signal Processing Techniques

Acoustic emission signals are sound waves generated in solid media which are similar to the sound waves propagated in air and other fluids but are more complex. The signal is affected by characteristics of the source, the path taken from the source to the sensor, the sensor's characteristics, and the measuring system. Generally the AE signals are intricate and using them to characterize the source could be difficult. Information is extracted by methods ranging from simple waveform parameter measurements to artificial intelligence (pattern recognition) approaches. The former often suffices for simple tests. The latter may be required for on-line monitoring of complex systems. Interpretation of the signals generated during the process often required advanced signal processing.

Both time domain and frequency domain analysis have certain limitations. A more sophisticated approach to machining process monitoring is to analyse the time frequency spectrum of the signal for patterns corresponding to the process characteristics of interest. The ability to detect these time-frequency characteristics is important, since the distribution of frequencies provides information about processing status. This approach has a potential advantage of insensitivity to signal intensity variations due to noise or other similar problems. By analyzing the time-frequency spectrum of signals, a larger amount of information can be extracted than investigating frequency spectrum only. Wavelet analysis is one such time-frequency analysis method.

According to the Heisenberg Uncertainty Principle, the product of signal dispersion f and its Fourier Transform dispersion $f$ is always greater than a constant c, which does not depend on the signal but only on the dimension of the space. Therefore, it is impossible to reduce both time and frequency localization arbitrarily. Wavelet analysis provides an interesting compromise on this problem. Applying windows with different sizes can change the resolution of time and frequency. Wavelet analysis allows the use of long intervals when more precise and low frequency information is needed, or the use of shorter intervals when high frequency information is needed.

A wavelet is a waveform with effectively limited duration and zero average value. If given wavelet function is,

$$\psi(t) \in L^2(R),$$

$$\psi_{ab}(t) = |a|^{\frac{1}{2}} \psi\left(\frac{t-b}{a}\right)$$

Where, a is $a$ scaling factor, $b$ is a shifting factor, and $L^2(R)$ is the set of signals of finite energy.

The continuous wavelet transform (CWT) is defined as the sum over all time of the signal multiplied by scaled, shifted versions of the wavelet function.

$$\left(W_\psi f\right)(a,b) = |a|^{\frac{1}{2}} \int_R f(t)\, \overline{\psi}\left(\frac{t-b}{a}\right) dt$$

Where, $a, b \in R, a \neq 0$, $R$ is the set of real numbers, and $\overline{\psi}(t)$ is conjugate function of $\psi(t)$

Calculating wavelet coefficients at every possible scale is arduous and time consuming. If scales and positions are based on powers of two, the analysis will be much more efficient. A fast wavelet decomposition and reconstruction algorithm was developed by Mallat in 1988. Mallat's algorithm for discrete wavelet transform (DWT) is a classical scheme in the signal processing community. It is well known as a twochannel sub-band coder, using conjugate quadrature filters or quadrature mirror filters (QMF). This very practical filtering algorithm yields a fast wavelet transform.

One of the fundamental relations of an orthogonal wavelet is the twin-scale relation (dilation equation or refinement equation) described by the following equation.

$$\phi(t) = \sqrt{2} \sum_{n \in Z} h_n \phi(2t - n)$$

Where $\phi(t)$ called scaling function and associated with wavelet function $\psi(t)$, and Z is the set of integer numbers.

All the filters used in DWT and inverse DWT (IDWT) are related to the sequence $(h_n)_{n \in Z}$ s compactly supported, the sequence $(h_n)$ is finite and can be viewed as a filter h. The filter h, known as the scaling filter, is a low-pass filter with length of 2N, sum of $\sqrt{2}$ and norm of 1. Written with the inner product:

$$h_k = <\phi_{j+1,0}, \ \phi_{j,k}>$$
$$g_k = <\psi_{j+1,0}, \ \phi_{j,k}>$$

Where, g and h are quadrature mirror filters.

Given a discrete sampling series $f(n)(n = 1, \wedge, N)$ of signal f(t) and denoting the approximation vector of signal at scale j = 0 as $C_0(n) = C_j(n)\big|_{j=0} = f(n)$, the dyadic discrete wavelet transform can be expressed as,

$$\left. \begin{aligned} C_{j+1}(n) &= \sum_{k \in Z} \overline{h}(k - 2n) C_j(k) \\ D_{j+1}(n) &= \sum_{k \in Z} \overline{g}(k - 2n) C_j(k) \end{aligned} \right]$$

The wavelet decomposition of signal f(t) can be written as,

$$A_j f(t) = A_{j+1} f(t) + D_{j+1} f(t)$$
$$= \sum_n C_{j+1}(n) \phi_{j+1,n}(t) + \sum_n D_{j+1}(n) \psi_{j+1,n}(t)$$

$A_j f$ is the output of applying a low-pass filter to f(t) and is called the approximation. $D_j f$ is the output of applying a series of width-variable band-pass filters to f(t) and is called the detail.

The key point of wavelet analysis is to extract information from the original signal by decomposing it into a series of approximations and details distributed over different frequency bands. The characteristics of frequency domain and time domain are preserved simultaneously. Further processing is then carried out after selecting several decomposition sequences suitable for the given application. The decomposition can be described as a decomposition tree shown in figure below.

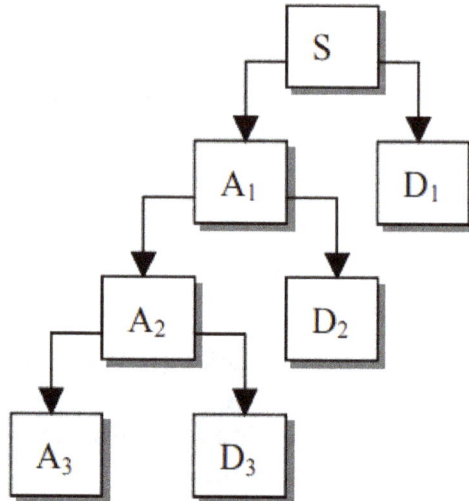

Wavelet decomposition tree

Based on the relation of frequency structure of wavelet decomposition, the frequency bandwidth of approximation and detail of level. $\left[ 0, \frac{1}{2} f_s \, 2^{-l} \right]$ and $\left[ \frac{1}{2} f_s \, 2^{-l}, \frac{1}{2} f_s \, 2^{-(l-1)} \right]$ respectively. It is noticed that the frequency band of every level is decomposed into two equal sub bands, the detail and the approximation. The result of wavelet translation is a series of decomposed signals belonging to different frequency bands.

## Cutter Location

A cutter location (CL) Data refers to the position to which a CNC milling machine has been instructed to hold a milling cutter by the G-code instructions. A neutral language file that helps to transfer instructions from CAM to a CNC machine. Each line of motion controlling G-code consists of two parts, first, the type of motion from the last cutter location to the next cutter location, e.g. "G01" means linear, "G02" means circular, and the next, the cutter location itself, the Cartesian point (20, 1.3, 4.409), e.g. "G01 X20 Y1.3 Z4.409".

The fundamental basis for creating the cutter paths suitable for Computer- Aided Manufacturing are functions that can find valid cutter locations, and stringing them together in a series. There are two broad and conflicting approaches to the problem of generating valid cutter locations, in a CAD model and a tool definition. The contents of the cutter locations are:

- Cutter location by offsets

- Cutter location against triangles

- Z Map

## Cutter Location by Offsets

It starts with a UV parametric point in a free form surface, the xyz point and the normal are calculated, and offset is taken from the point along the normal in a way consistent with the tool definition so that the cutter is now tangent to the surface at that point. It may collide with the model elsewhere and there is no indication to this happening until the full implementation of the triangulated approach.

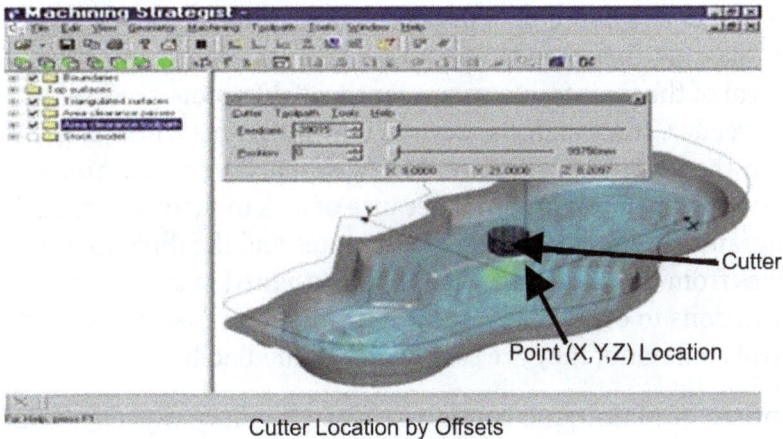

Cutter Location by Offsets

## Cutter Location Against Triangles

It starts with the XY component for a cutter location and loop across every triangle in the model. For each triangle which crosses under the circular shadow of the cutter, the Z value of the cutter location required for it is calculated to exactly touch the triangle, and find the maximum of all such values.

## Problems

It requires a lot of memory to hold enough triangles to register the model at a tight enough tolerance, and it takes longer to program to get the initial cutter location values. However, they are at least guaranteed in all cases. This is how all major CAM systems do it these days because it works without failing, no matter what the complexity and geometry of the model may be. Reliability is far more important than efficiency. The above refers only to 3-axis machines and 5-axis machines need a special entry of their own.

## Z Map

The Z Map algorithm is a regular array of Cutter Location values in the computer memory. The result is a model of the height map of cutter positions from which the values can be interpolated. Due to accuracy issues, this was generalized into an Extended Z Map, by the placement of "floating" points between the fixed Z Map points. The locations of the Extended Z Map points are found when the Z Map is created. Extended Z Map points are only placed where sharp edges occur between the normal Z Map points, completely flat source geometry will not require any Extended Z Map points.

## Direct Numerical Control

Direct Numerical Control (DNC) is a process set in a manufacturing unit where a set of machines are controlled by a programmed computer with the help of a direct connection to the same. It is based on real-time data and involves data collection from the machines and passing the same to the mainframe, at regular intervals. The operator will be in control of the mainframe computer through remote accesses. A DNC will not contain a tape reader. Instead, it has several part programs which is transferred to the machines from the computer memory. In few scenarios the machine controllers will not be able to store the entire program because of lack of memory space. In such situations, the program is stored in a different computer and the directions are sent directly to the machines from that location. The DNC is designed in such a way that it provides separate instructions to every machine on the system. In case, where the machine immediate control command, they are sent across immediately.

A DNC comprises of Mainframe computer, Huge memory capacity, Connectivity between the machines and the computers, and Machine tools.

Direct numerical control should only be used as a last resort. If you're running long programs on a regular basis, you should choose a CNC that is designed for running long programs. Many high speed machining controls run programs from a hard drive within the control. Program length is limited only by hard drive capacity.

However, if you only occasionally run long programs—or if you have a machine that cannot run long programs from internal memory—you may have no alternative but to use direct numerical control. Let's look at how this system is used.

With most current CNCs, the same connection used to transfer programs is used for direct numerical control, especially if there is only one communications port on the machine. So if you currently have the ability to transfer programs, you won't have to change any communications protocol in order to perform direct numerical control.

Most controls that can perform direct numerical control have a special parameter to specify that DNC will be used. With one popular control model, this parameter is named remote program execution (simply remote on the display screen setting page).

When this setting parameter is enabled, the control will execute programs coming in to the control from the communications port instead of from internal memory. In either case, the mode switch will still be placed in the auto or memory position when the program is activated.Assuming the setting parameter for remote program execution is enabled, actually running a program from the communications port is pretty simple. This procedure usually requires the CNC operator to send the program from the DNC system to the machine in the same way he or she would normally send a program into the machine's memory. The operator may have to leave the machine to do so.

For this reason, we recommend starting by turning on the single block switch. This will ensure that only one command will be activated (commonly the program number command) when the machine starts running the program. This is especially important if the operator will not be close to the machine when program execution begins.

Next, the operator will press the cycle start button. This will cause the control to look to the communications port for the program. In essence, it's like pressing the "read" button when loading a program using a distributive numerical control system. As soon as the control sees a program coming through the communications port, it will begin executing it—which is why it is important to have the single block switch turned on.

Finally, the operator must command the DNC system to send the program to the machine. As stated, when the machine receives the program, it will begin executing it. All safety functions will still be available, so the operator can manipulate the way the machine will execute the program.

Most CNC controls provide a way to restart a program beginning from a specified command. In normal operation, the program restart function is not commonly needed, but it is especially important with direct numerical control.

When executing a program from within the machine's memory, the operator can simply scan within the program (in the edit mode) to the restart command. When they execute the program, the machine will begin executing from the restart command.

## Types of DNC

- RS232 – based DNC system: Operates with a switch box or multi-port cable connections to connect several machines.

- Terminal – based DNC system: A CNC terminal is created to connect several programs.

- Network – based system: Direct access to NC programs without installation of terminals and is thus highly efficient.

## Advantages

- It avoids the usage of punched taps and the reader from the system

- Helps the business to understand production performance by getting several reports and useful data from the machines

- Helps in building a centralized control for the machines

- Useful for time management and increased productivity

- Convenient storage of part programs in several computer files.

# Master Production Schedule

MPS is the process that helps manufacturers plan which products and related quantities to produce during certain periods. MPS is proactive in that it actually drives the production process in terms of what is manufactured and what materials are procured. MPS also serves a bridge to sales as it informs them about what is available to promise to customers and when deliveries can be made.

## MPS as a Crucial Planning Function

As part of a fully integrated ERP system, MPS typically provides a crucial planning function, extracting actual supply and demand data, as well as forecasts, to deliver accurate and timely production plans that help manufacturers achieve their production objectives and minimize procurement costs. MPS also takes into account the manufacturing capacity of the plant in its calculations. Once production orders have been analyzed and approved, the MRP process is initiated and purchase orders can be generated. MPS also provides protection against shortages, unexpected scheduling snafus and inefficient allocation of resources.

## MPS and MRP – their Relationship

While MPS and MRP have some similarities, including the ability to generate planned manufacture, purchase and transfer orders, there are a few characteristics that set them apart. For example, unlike MRP, MPS tends to focus its planning capabilities on the production of finished goods, components or parts that generate the greatest profitability for the manufacturer and are therefore likely to constitute the lion's share of the resources needed for production.

Another difference is that the MPS operates only within one level of an item's BOM, while MRP can be utilized at every level. Also, MRP focuses its planning capabilities more toward meeting demand for component parts or subassemblies, while MPS focuses more on establishing production plans to satisfy the actual demand for finished products, as well as to meet projected customer delivery dates.

MPS plans are typically based upon input such as actual sales orders, service orders, available resources, inventory levels, capacity constraints or forecasts. These forecasts give manufacturers the ability to anticipate product demand and the flexibility to adjust production plans based on seasonality, promotions and fluctuating demand for particular items and/or finished products. The output from MPS includes quantities of an item to be produced, due dates and quantities available to promise.

## Benefits of MPS

- Ability to make adjustments to fluctuations in demand and still minimize waste
- Helps prevent shortages and scheduling mishaps
- Improves efficiency in the location of production resources
- Provides more effective cost controls and more accurate estimates of material requirements and delivery dates
- Reduces lead times throughout the year
- Provides an effective communication conduit with the sales team for planning purposes.

## MPS: Heart of a Manufacturing ERP System

MPS lies at the heart of a manufacturing ERP system, and connects to multiple modules including Accounts Payable (AP), Accounts Receivable (AR), General Ledger (GL), Customer Relationship Management (CRM), Inventory, Purchasing, Lot Tracking and Human Resources (HR), just to name a few.

## Common MPS Reports

Some of the standard reports created by a MPS include:

- Available-to-Promise −Presents the available-to-promise quantities for each MPS line item. The report is time-phased, usually into weekly or monthly "buckets."
- Demand Tracking Report − Provides historical data on actual shipments and order bookings as compared to management forecasts.
- Forecast Data Report − Summary of historical demand activity, which indicates the significance of errors between forecast and actual, and provides a statistical summary.

- Period Summary Forecast – Forecast by line item within a product group for each period through the future 12 periods, with summaries by period for the group and yearly for line items.

- Item Demand and Forecast – Presents several years of historical data (user-specified) and the next 12 months of forecast demand for each item. Typical data elements can include YTD totals, total yearly demand and quarterly totals, with comparisons by percent between items and their total product group.

- Build Schedule Report – Reports the build schedule for one or all assemblies.

- Schedule versus Actual Output – Reports the actual output compared with the scheduled output at a particular work center.

- Where Used Report – Lists all parts/tools used at each work center/machine.

## Example

In our example we have just a single product being produced.

Production takes place each period (week) either in the normal (regular) production shift or in overtime associated with that shift. There is only one shift (i.e. not operating a two/three shift system - such as with "round-the-clock" working).

Completed items can also be "bought-in" from a subcontractor (at a cost).

We are allowed to hire/fire workers (again at a cost). Backorders are also allowed (recall here that backorders are customer orders that cannot be satisfied in the required period, but the customer allows the order to remain open to be fulfilled in a later period). Lost sales are not allowed.

The diagram below illustrates the situation and the types of factor with which we are dealing graphically.

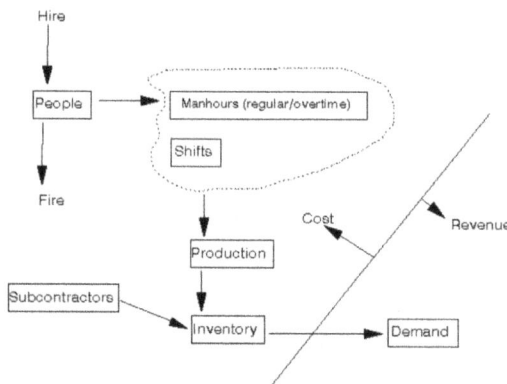

The data for the example we consider is as below, where we have shown the initial data entry screen from the package.

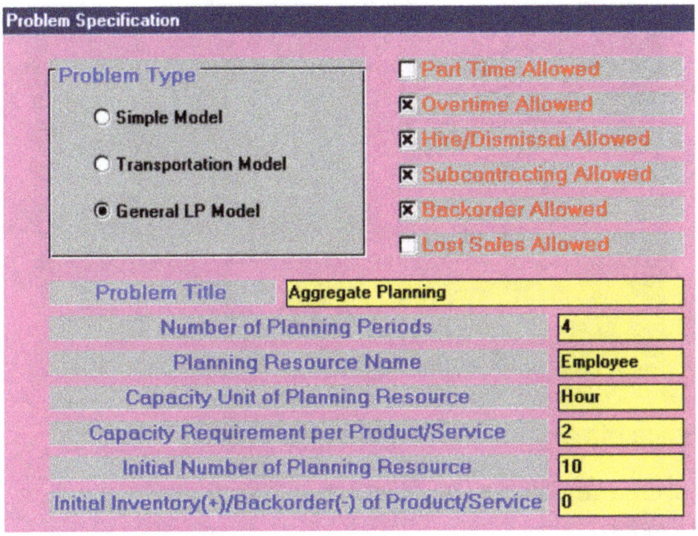

**Problem Specification**

Problem Type
- ○ Simple Model
- ○ Transportation Model
- ◉ General LP Model

- ☐ Part Time Allowed
- ☒ Overtime Allowed
- ☒ Hire/Dismissal Allowed
- ☒ Subcontracting Allowed
- ☒ Backorder Allowed
- ☐ Lost Sales Allowed

| | |
|---|---|
| Problem Title | Aggregate Planning |
| Number of Planning Periods | 4 |
| Planning Resource Name | Employee |
| Capacity Unit of Planning Resource | Hour |
| Capacity Requirement per Product/Service | 2 |
| Initial Number of Planning Resource | 10 |
| Initial Inventory(+)/Backorder(-) of Product/Service | 0 |

In the above screen we have chosen the "General LP Model". This is the most general of the options allowed by the package. LP stands for linear programming and is a generalized way of modeling decision problems. To ease data entry we have not crossed the "Part Time Allowed" box - if we had then we would have had the option of dealing with part time employees.

We have also not crossed the "Lost Sales Allowed" box - if we had then we would have allowed lost sales. In general a company may allow lost sales because the company finds that customers simply do not backorder - i.e. a lost sale is automatic if the product is not immediately available; or the company is prepared to allow lost sales as it may be better to allow orders to be lost than to allow such orders to become backorders (thereby incurring backorder costs).

The remaining boxes have been crossed and so we can deal with:

- Overtime
- Hiring/firing
- Subcontracting
- Backorders

In our example above we have just 4 periods (weeks) - this is our time horizon (planning period). We are dealing with employees working hours in each week. Two employee hours are required to produce one unit of each product and the initial number of employees is 10. At the start of the planning period there is no initial inventory (nor are there any backorders).

The data for our example entered into the package in the light of the choices made at the initial screen is as below:

| DATA ITEM | Period 1 | Period 2 | Period 3 | Period 4 |
|---|---|---|---|---|
| Forecast Demand | 250 | 340 | 400 | 450 |
| Initial Number of Employee | 10 | 10 | 10 | 10 |
| Regular Time Capacity in Hour per Employee | 35 | 35 | 35 | 35 |
| Regular Time Cost per Hour | 15 | 15 | 15 | 15 |
| Undertime Cost per Hour | | | | |
| Overtime Capacity in Hour per Employee | 10 | 10 | 10 | 10 |
| Overtime Cost per Hour | 25 | 25 | 25 | 25 |
| Hiring Cost per Employee | 500 | 500 | 500 | 500 |
| Dismissal Cost per Employee | 2000 | 2000 | 2000 | 2000 |
| Maximum Number of Employee Allowed | M | M | M | M |
| Minimum Number of Employee Allowed | 8 | 8 | 8 | 8 |
| Initial Inventory (+) or Backorder (-) | | | | |
| Maximum Ending Inventory Allowed | M | M | M | M |
| Minimum Ending Inventory (Safety Stock) | | | | |
| Unit Inventory Holding Cost | 3 | 3 | 3 | 3 |
| Maximum Subcontracting Allowed | M | M | M | M |
| Unit Subcontracting Cost | 60 | 60 | 60 | 60 |
| Maximum Backorder Allowed | M | M | M | M |
| Unit Backorder Cost | M | M | M | M |
| Other Unit Production Cost | | | | |
| Capacity Requirement in Hour per Unit | 2 | 2 | 2 | 2 |

The meaning of each of these lines of data is given below:

- Forecast Demand- this is the forecast demand for the product in each of our 4 periods (weeks).

- Initial Number of Employee- this is the initial number of employees in each week, here just the 10 employees we have currently.

- Regular Time Capacity in Hour per Employee- this is how many regular hours each employee works per week, here 35 hours.

- Regular Time Cost per Hour- this is the cost per hour of regular time worked, here £15.

- Under time Cost per Hour- this is the cost per hour of not using a worker to their full regular capacity, here zero.

- Overtime Capacity in Hour per Employee- this is the maximum number of hours each employee can work in overtime per week, here 10 hours.

- Overtime Cost per Hour- this is the cost per hour of overtime, here £25.

- Hiring Cost per Employee- this is the cost of hiring one employee, here £500.

- Dismissal Cost per Employee- this is the cost of dismissing (firing) one employee, here £2000.

- Maximum/Minimum Number of Employee Allowed- here we can set limits on the maximum and minimum number of employees, here M signifies there is no limit on the maximum number and the minimum number is 8. In general there may be an upper limit on the number of employees due to physical capacity constraints.

- Initial Inventory (+) or Backorder (-)- the initial inventory available or backorders outstanding, here zero.

- Maximum/Minimum Ending Inventor- here we can set limits on the maximum and minimum number of product units in stock at the end of each week, here M signifies there is no limit on the maximum number and the minimum number is zero. In general there may be an upper limit because we have a limited space in which to store stock. The minimum number corresponds to safety stock that may be kept in case of unforeseen demand.

- Unit Inventory Holding Cost- this is the cost of holding one unit in stock at the end of each period, here £3.

- Maximum Subcontracting Allowed- this is the maximum number of product units we are allowed to buy in from the external subcontractor, here there is no limit on the amount that may be bought in. In general there may be a limit on the total amount the subcontractor can supply to us each period.

- Unit Subcontracting Cost- this is the cost of each unit bought from the external subcontractor, here £60.

- Maximum Backorder Allowed- this is the maximum number of backorders allowed at the end of each period, here there is no limit on the number of backorders that can be held at the end of each period.

- Unit Backorder Cos - this is the cost of each backorder outstanding at the end of each period, here the M signifies that each backorder is very expensive. The effect of M here will be to ensure that (if at all possible) backorders will be avoided.

- Other Unit Production Cost- this is the cost of producing one unit of the product that is not already accounted for by employee costs - here zero.

- Capacity Requirement in Hour per Unit- this is the number of employee hours that are required to produce one unit of the product, here 2 hours.

In order to ease understanding of the problem most of the above data items take the same value in each and every period (week). However it would be perfectly possible for them to have different values in each week.

## Milling

Milling is a process performed with a machine in which the cutters rotate to remove the material from the work piece present in the direction of the angle with the tool axis. With the help of the milling machines one can perform many operations and functions starting from small objects to large ones.

Milling machining is one of the very common manufacturing processes used in machinery shops and industries to manufacture high precision products and parts in different shapes and sizes.

## Milling Machine

The milling machines are also known as the multi-tasking machines (MTMs) which are multi-purpose machines capable of milling and turning the materials as well. The milling machine has got the cutter installed up on it which helps in removing the material from the surface of the work piece. When the material gets cooled down then it is removed from the milling machine.

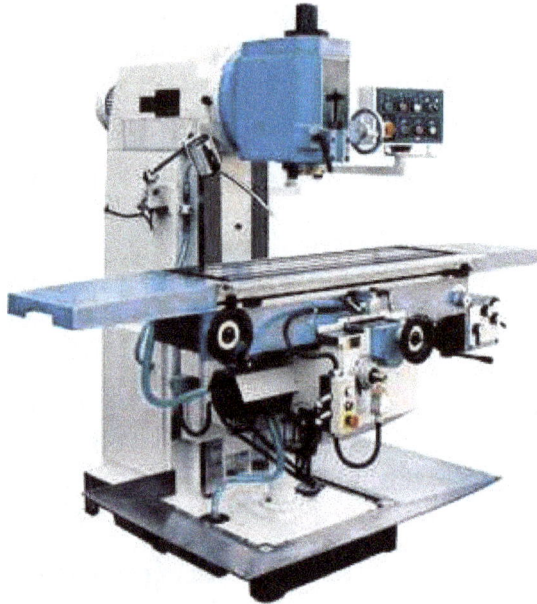

## Milling Process

The milling machine involves the following processes or phases of cutting.

## Milling Cutters

There are a lot of cutting tools used in the milling process. The milling cutters named end mills have special cutting surfaces on their end surfaces so that they can be placed onto the work piece by drilling. These also have extended cutting surfaces on each side for the purpose of peripheral milling. The milling cutters have small cutters at the end corners. The cutters are made from highly resistant materials that are durable and produce less friction.

## Surface Finish

Any material put through the cutting area of the milling machine gets regular intervals. The side cutters have got regular ridges on them. The distance between the ridges depends on the feed rate, the diameter of the cutter and the quantity of cutting surfaces. These can be the significant variations in the height of the surfaces.

## Gang Milling

This means that more than two milling cutters are involved in a setup like the horizontal milling. All the cutters perform a uniform operation or it may also be possible that the cutter may perform distinct operations. This is an important operation for producing duplicate parts.

## Equipment

Milling is performed with a milling cutter in various forms, held in a collet or similar which, in turn, is held in the spindle of a milling machine.

## Types and nomenclature

Mill orientation is the primary classification for milling machines. The two basic configurations are vertical and horizontal. However, there are alternative classifications according to method of control, size, purpose and power source.

## Mill Orientation

## Vertical Mill

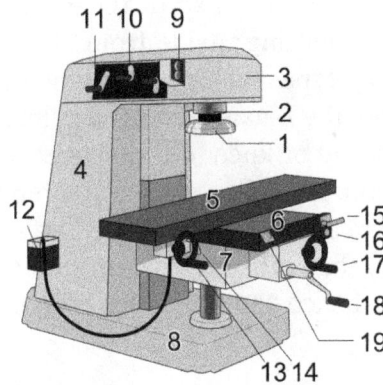

Vertical milling machine

In the vertical mill the spindle axis is vertically oriented. Milling cutters are held in the spindle and rotate on its axis. The spindle can generally be extended (or the table can be raised/lowered, giving the same effect), allowing plunge cuts and drilling. There are two subcategories of vertical mills: the bed mill and the turret mill.

A turret mill has a stationary spindle and the table is moved both perpendicular and parallel to the spindle axis to accomplish cutting. The most common example of this type is the Bridgeport, described below. Turret mills often have a quill which allows the milling cutter to be raised and lowered in a manner similar to a drill press. This type of machine provides two methods of cutting in the vertical (Z) direction: by raising or lowering the quill, and by moving the knee.

In the bed mill, however, the table moves only perpendicular to the spindle's axis, while the spindle itself moves parallel to its own axis.

Turret mills are generally considered by some to be more versatile of the two designs. However, turret mills are only practical as long as the machine remains relatively small. As machine size increases, moving the knee up and down requires considerable effort and it also becomes difficult to reach the quill feed handle (if equipped). Therefore, larger milling machines are usually of the bed type.

A third type also exists, a lighter machine, called a mill-drill, which is a close relative of the vertical mill and quite popular with hobbyists. A mill-drill is similar in basic configuration to a small drill press, but equipped with an X-Y table. They also typically use more powerful motors than a comparably sized drill press, with potentiometer-controlled speed and generally have more heavy-duty spindle bearings than a drill press to deal with the lateral loading on the spindle that is created by a milling operation. A mill drill also typically raises and lowers the entire head, including motor, often on a dovetailed vertical, where a drill press motor remains stationary, while the arbor raises and lowers within a driving collar.

Other differences that separate a mill-drill from a drill press may be a fine tuning adjustment for the Z-axis, a more precise depth stop, the capability to lock the X, Y or Z axis, and often a system of tilting the head or the entire vertical column and power head assembly to allow angled cutting. Aside from size and precision, the principal difference between these hobby-type machines and larger true vertical mills is that the X-Y table is at a fixed elevation; the Z-axis is controlled in basically the same fashion as drill press, where a larger vertical or knee mill has a vertically fixed milling head, and changes the X-Y table elevation. As well, a mill-drill often uses a standard drill press-type Jacob's chuck, rather than an internally tapered arbor that accepts collets. These are frequently of lower quality than other types of machines, but still fill the hobby role well because they tend to be bench top machines with small footprints and modest price tags.

## Horizontal Milling Machine

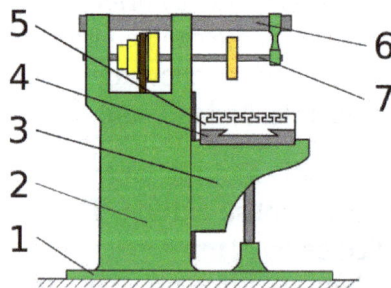

Horizontal milling machine. 1: base 2: column 3: knee 4 & 5: table (x-axis slide is integral) 6: overarm 7: arbor (attached to spindle)

A horizontal mill has the same sort but the cutters are mounted on a horizontal spindle across the table. Many horizontal mills also feature a built-in rotary table that allows milling at various angles; this feature is called a universal table. While end mills and the other types of tools available to a vertical mill may be used in a horizontal mill, their real advantage lies in arbor-mounted cutters, called side and face mills, which have a cross section rather like a circular saw, but are generally wider and smaller in diameter. Because the cutters have good support from the arbor and have a larger cross-sectional area than an end mill, quite heavy cuts can be taken enabling rapid material removal rates. These are used to mill grooves and slots. Plain mills are used to shape flat surfaces. Several cutters may be ganged together on the arbor to mill a complex shape of slots and planes. Special cutters can also cut grooves, bevels, radii, or indeed any section desired. These specialty cutters tend to be expensive. Simplex mills have one spindle, and duplex mills have two. It is also easier to cut gears on a horizontal mill. Some horizontal milling machines are equipped with a power-take-off provision on the table. This allows the table feed to be synchronized to a rotary fixture, enabling the milling of spiral features such as hypoid gears.

## Comparative Merits

The choice between vertical and horizontal spindle orientation in milling machine design usually hinges on the shape and size of a work piece and the number of sides of the work piece that require machining. Work in which the spindle's axial movement is normal to one plane, with an end mill as the cutter, lends itself to a vertical mill, where the operator can stand before the machine and have easy access to the cutting action by looking down upon it. Thus vertical mills are most favored for die sinking work (machining a mould into a block of metal). Heavier and longer work pieces lend themselves to placement on the table of a horizontal mill.

Prior to numerical control, horizontal milling machines evolved first, because they evolved by putting milling tables under lathe-like headstocks. Vertical mills appeared in subsequent decades, and accessories in the form of add-on heads to change horizontal mills to vertical mills (and later vice versa) have been commonly used. Even in the CNC era, a heavy work piece needing machining on multiple sides lends itself to a horizontal machining center, while die sinking lends itself to a vertical one.

## Alternative Classifications

In addition to horizontal versus vertical, other distinctions are also important:

| Criterion | Example classification scheme | Comments |
|---|---|---|
| Spindle axis orientation | Vertical versus horizontal; Turret versus non-turret | Among vertical mills, "Bridgeport-style" is a whole class of mills inspired by the Bridgeport original, rather like the IBM PC spawned the industry of IBM-compatible PCs by other brands |

| Control | Manual; Mechanically automated via cams; Digitally automated via NC/CNC | In the CNC era, a very basic distinction is manual versus CNC. Among manual machines, a worthwhile distinction is non-DRO-equipped versus DRO-equipped |
|---|---|---|
| Control (specifically among CNC machines) | Number of axes (e.g., 3-axis, 4-axis, or more) | Within this scheme, also:<br><br>• Pallet-changing versus non-pallet-changing<br><br>• Full-auto tool-changing versus semi-auto or manual tool-changing |
| Purpose | General-purpose versus special-purpose or single-purpose | |
| Purpose | Tool room machine versus production machine | Overlaps with above |
| Purpose | "Plain" versus "universal" | A distinction whose meaning evolved over decades as technology progressed, and overlaps with other purpose classifications above. Not relevant to today's CNC mills. Regarding manual mills, the common theme is that "plain" mills were production machines with fewer axes than "universal" mills; for example, whereas a plain mill had no indexing head and a non-rotating table, a universal mill would have those. Thus it was suited to universal service, that is, a wider range of possible tool paths. Machine tool builders no longer use the "plain"-versus-"universal" labeling. |
| Size | Micro, mini, bench top, standing on floor, large, very large, gigantic | |
| Power source | Line-shaft-drive versus individual electric motor drive | Most line-shaft-drive machines, ubiquitous circa 1880–1930, have been scrapped by now |
| | Hand-crank-power versus electric | Hand-cranked not used in industry but suitable for hobbyist micro mills |

## Variants

- Bed mill: This refers to any milling machine where the spindle is on a pendant that moves up and down to move the cutter into the work, while the table sits on a stout bed that rests on the floor. These are generally more rigid than a knee mill. Gantry mills can be included in this bed mill category.

- Box mill or column mill: Very basic hobbyist bench-mounted milling machines that feature a head riding up and down on a column or box way.

- C-frame mill: These are larger, industrial production mills. They feature a knee and fixed spindle head that is only mobile vertically. They are typically much more powerful than a turret mill, featuring a separate hydraulic motor for integral hydraulic power feeds in all directions, and a twenty to fifty horsepower motor. Backlash eliminators are almost always standard equipment. They use large NMTB 40 or 50 tooling. The tables on C-frame mills are usually 18" by 68" or larger, to allow multiple parts to be machined at the same time.

A Sieg  X2 miniature hobbyist mill plainly showing the basic parts of a mill

- Floor mill: These have a row of rotary tables, and a horizontal pendant spindle mounted on a set of tracks that runs parallel to the table row. These mills have predominantly been converted to CNC, but some can still be found (if one can even find a used machine available) under manual control. The spindle carriage moves to each individual table, performs the machining operations, and moves to the next table while the previous table is being set up for the next operation. Unlike other mills, floor mills have movable floor units. A crane drops massive rotary tables, X-Y tables, etc., into position for machining, allowing large and complex custom milling operations.

- Gantry mill: The milling head rides over two rails (often steel shafts) which lie at each side of the work surface.

- Horizontal boring mill: Large, accurate bed horizontal mills that incorporate many features from various machine tools. They are predominantly used to create large manufacturing jigs, or to modify large, high precision parts. They have a spindle stroke of several (usually between four and six) feet, and many are equipped with a tailstock to perform very long boring operations without losing accuracy as the bore increases in depth. A typical bed has X and Y travel, and is between three and four feet square with a rotary table or a larger rectangle without a table. The pendant usually provides between four and eight feet of vertical movement. Some mills have a large (30" or more) integral facing head. Right angle rotary tables and vertical milling attachments are available for further flexibility.

- Jig borer: Vertical mills that are built to bore holes, and very light slot or face milling. They are typically bed mills with a long spindle throw. The beds are more accurate, and the hand wheels are graduated down to .0001" for precise hole placement.

- Knee mill or knee-and-column mill: refers to any milling machine whose x-y table rides up and down the column on a vertically adjustable knee. This includes Bridge ports.

- Planer-style mill Large mills built in the same configuration as planers except with a milling spindle instead of a planing head. This term is growing dated as planers themselves are largely a thing of the past.

- Ram-type mill: This can refer to any mill that has a cutting head mounted on a sliding ram. The spindle can be oriented either vertically or horizontally. In practice most mills with rams also involve swiveling ability, whether or not it is called "turret" mounting. The Bridgeport configuration can be classified as a vertical-head ram-type mill. Van Norman specialized in ram-type mills through most of the 20th century. Since the wide dissemination of CNC machines, ram-type mills are still made in the Bridgeport configuration (with either manual or CNC control), but the less common variations (such as were built by Van Norman, Index, and others) have died out, their work being done now by either Bridgeport-form mills or machining centers.

- Turret mill:  More commonly referred to as Bridgeport-type milling machines. The spindle can be aligned in many different positions for a very versatile, if somewhat less rigid machine.

## Alternative Terminology

A milling machine is often called a mill by machinists. The archaic term miller was commonly used in the 19th and early 20th centuries.

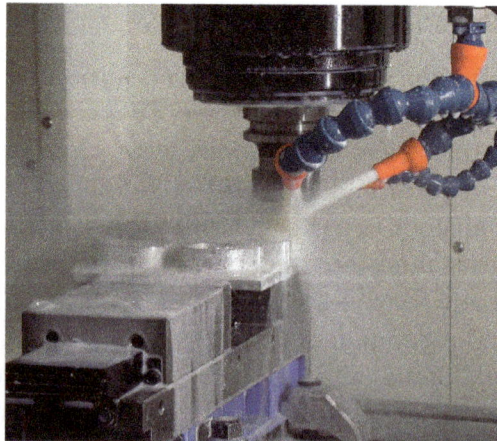

Thin wall milling of aluminum using a water based cutting fluid on the milling cutter

Since the 1960s there has developed an overlap of usage between the terms milling machine and machining center. NC/CNC machining centers evolved from milling machines, which is why the terminology evolved gradually with considerable overlap that still persists. The distinction, when one is made, is that a machining center is a mill with features that pre-CNC mills never had, especially an automatic tool changer (ATC) that includes a tool magazine (carousel), and sometimes an automatic pallet changer (APC). In typical usage, all machining centers are mills, but not all mills are machining centers; only mills with ATCs are machining centers.

Most CNC milling machines (also called machining centers) are computer controlled vertical mills with the ability to move the spindle vertically along the Z-axis. This extra degree of freedom permits their use in die sinking, engraving applications, and 2.5D surfaces such as relief sculptures. When combined with the use of conical tools or a ball nose cutter, it also significantly improves milling precision without impacting speed, providing a cost-efficient alternative to most flat-surface hand-engraving work.

Five-axis machining center with rotating table and computer interface

CNC machines can exist in virtually any of the forms of manual machinery, like horizontal mills. The most advanced CNC milling-machines, the multi axis machine, add two more axes in addition to the three normal axes (XYZ). Horizontal milling machines also have a C or Q axis, allowing the horizontally mounted work piece to be rotated, essentially allowing asymmetric and eccentric turning. The fifth axis (B axis) controls the tilt of the tool itself. When all of these axes are used in conjunction with each other, extremely complicated geometries, even organic geometries such as a human head can be made with relative ease with these machines. But the skill to program such geometries is beyond that of most operators. Therefore, 5-axis milling machines are practically always programmed with CAM.

The operating system of such machines is a closed loop system and functions on feedback. These machines have developed from the basic NC (Numeric Control) machines. A computerized form of NC machines is known as CNC machines. A set of instructions (called a program) is used to guide the machine for desired operations. Some very commonly used codes, which are used in the program are:

G00 – rapid traverse

G01 – linear interpolation of tool.

G21 – dimensions in metric units.

M03/M04 – spindle start (clockwise/counter clockwise).

T01 M06 – automatic tool change to tool 1

M30 – program end.

Various other codes are also used. A CNC machine is operated by a single operator called a programmer. This machine is capable of performing various operations automatically and economically.

With the declining price of computers and open source CNC software, the entry price of CNC machines has plummeted.

High speed steel with cobalt end mills used for cutting operations in a milling machine

## Tooling

The accessories and cutting tools used on machine tools (including milling machines) are referred to in aggregate by the mass noun "tooling". There is a high degree of standardization of the tooling used with CNC milling machines, and a lesser degree with manual milling machines. To ease up the organization of the tooling in CNC production many companies use a tool management solution.

Milling cutters for specific applications are held in various tooling configurations.

CNC milling machines nearly always use SK (or ISO), CAT, BT or HSK tooling. SK tooling is the most common in Europe, while CAT tooling, sometimes called V-Flange Tooling, is the oldest and probably most common type in the USA. CAT tooling was invented by Caterpillar Inc. of Peoria, Illinois, in order to standardize the tooling used on their machinery. CAT tooling comes in a range of sizes designated as CAT-30, CAT-40, CAT-50, etc. The number refers to the Association for Manufacturing Technology (formerly the National Machine Tool Builders Association (NMTB)) Taper size of the tool.

A CAT-40 tool holder

A boring head on a Morse taper shank

An improvement on CAT Tooling is BT Tooling, which looks similar and can easily be confused with CAT tooling. Like CAT Tooling, BT Tooling comes in a range of sizes and uses the same NMTB body taper. However, BT tooling is symmetrical about the spindle axis, which CAT tooling is not. This gives BT tooling greater stability and balance at high speeds. One other subtle difference between these two tool holders is the thread used to hold the pull stud. CAT Tooling is all Imperial thread and BT Tooling is all Metric thread. Note that this affects the pull stud only, it does not affect the tool that they can hold, both types of tooling are sold to accept both Imperial and metric sized tools.

SK and HSK tooling, sometimes called "Hollow Shank Tooling", is much more common in Europe where it was invented than it is in the United States. It is claimed that HSK tooling is even better than BT Tooling at high speeds. The holding mechanism for HSK tooling is placed within the (hollow) body of the tool and, as spindle speed increases, it expands, gripping the tool more tightly with increasing spindle speed. There is no pull stud with this type of tooling.

For manual milling machines, there is less standardization, because a greater plurality of formerly competing standards exist. Newer and larger manual machines usually use NMTB tooling. This tooling is somewhat similar to CAT tooling but requires a draw-bar within the milling machine. Furthermore, there are a number of variations with NMTB tooling that make interchangeability troublesome. The older a machine, the greater the plurality of standards that may apply (e.g., Morse, Jarno, Brown & Sharpe, Van Norman, and other less common builder-specific tapers). However, two standards that have seen especially wide usage are the Morse #2 and the R8, whose prevalence was driven by the popularity of the mills built by Bridgeport Machines of Bridgeport,

Connecticut. These mills so dominated the market for such a long time that "Bridge-port" is virtually synonymous with "manual milling machine". Most of the machines that Bridgeport made between 1938 and 1965 used a Morse taper #2, and from about 1965 onward most used an R8 taper.

## Accessories

- Arbor support
- Stop block

## CNC Pocket Milling

Pocket milling has been regarded as one of the most widely used operations in machining. It is extensively used in aerospace and shipyard industries. In pocket milling the material inside an arbitrarily closed boundary on a flat surface of a work piece is removed to a fixed depth. Generally flat bottom end mills are used for pocket milling. Firstly roughing operation is done to remove the bulk of material and then the pocket is finished by a finish end mill. Most of the industrial milling operations can be taken care of by 2.5 axis CNC milling. This type of path control can machine up to 80% of all mechanical parts. Since the importance of pocket milling is very relevant, therefore effective pocketing approaches can result in reduction in machining time and cost. NC pocket milling can be carried out mainly by two tool paths, viz. linear and non-linear.

## Linear Tool Path

In this approach, the tool movement is unidirectional. Zig-zag and zig tool paths are the examples of linear tool path.

## Zig-Zag Tool Path

In zig-zag milling, material is removed both in forward and backward paths. In this case, cutting is done both with and against the rotation of the spindle. This reduces the machining time but increases machine chatter and tool wear.

## Zig Tool Path

In zig milling, the tool moves only in one direction. The tool has to be lifted and re-tracted after each cut, due to which machining time increases. However, in case of zig milling surface quality is better.

## Non-linear Tool Path

In this approach, tool movement is multi-directional. One example of non-linear tool path is contour-parallel tool path.

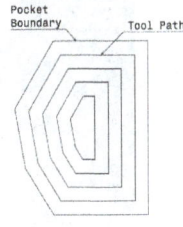

## Contour-parallel Tool Path

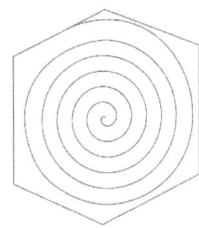

In this approach, the required pocket boundary is used to derive the tool path. In this case the cutter is always in contact with the work material. Hence the idle time spent in positioning and retracting the tool is avoided. For large-scale material removal, contour-parallel tool path is widely used because it can be consistently used with up-cut or down-cut method during the entire process. There are three different approaches that fall into the category of contour-parallel tool path generation. They are:

- Pair-wise intersection approach: In pair-wise intersection approach, the boundary of the pocket is brought inwards in steps, The offset segments will intersect at concave corners. To obtain the required contour, these intersections are to be trimmed off. On the other hand, in case of convex corner, the offset segments are extended and thereby connected to make the contour. These operations viz. offsetting, trimming and extending are repeatedly done to cover the entire machining volume with sufficient layer of profiles.

- Voronoi diagram approach: In voronoi diagram approach, the pocket boundary is segmented and voronoi diagram is constructed for the entire pocket boundary. These voronoi diagrams are used for generating the tool path for machining. This method is considered to be more efficient and robust. Moreover, it avoids topological problems associated with traditional offsetting algorithms.

## Curvilinear Tool Path

In this approach, the tool travels along a gradually evolving spiral path. The spiral starts at the center of the pocket to be machined and the tool gradually moves towards the pocket boundary. The direction of the tool path changes progressively and local acceleration and deceleration of the tool are minimized. This reduces tool wear.

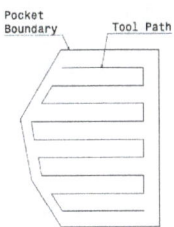

| Zig-zag tool path | Zig tool path | Contour-parallel tool path | Curvilinear tool path |

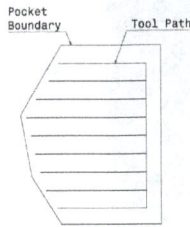

## Multi-axis Machining

Multi-axis machining offers some serious capability advantages to SME Manufacturers. Multi-axis machines are capable of producing highly complicated components by moving the table (and work piece) as well as the machine tool. This combination provides the ability move in multiple axes, over and above the 3 (X,Y and Z) typical of conventional milling machines.

## 5-axis CNC Machining

In the simplest terms, 5-axis machining involves using a CNC to move a part or cutting tool along five different axes simultaneously. This enables the machining of very complex parts, which is why 5-axis is especially popular for aerospace applications.

However, several factors have contributed to the wider adoption of 5-axis machining. These include:

- A push toward single-setup machining (sometimes referred to as "Done-in-One") to reduce lead time and increase efficiency.

- The ability to avoid collision with the tool holder by tilting the cutting tool or the table, which also allows better access to part geometry.

- Improved tool life and cycle time as a result of tilting the tool/table to maintain optimum cutting position and constant chip load.

## Axes in 5-Axis

We all know the story about Newton and the apple, but there's a similarly apocryphal story about the mathematician and philosopher, Rene Descartes.

Rene Descartes

Descartes was lying in bed (as mathematicians and philosophers are wont to do) when he noticed a fly buzzing around his room. He realized that he could describe the fly's position in the room's three-dimensional space using just three numbers, represented by the variables X, Y and Z.

This is the Cartesian Coordinate system, and it's still in use more than three centuries after Descartes' death. So X, Y and Z cover three of the five axes in 5-axis machining.

What about the other two?

Imagine zooming in on Descartes' fly in mid-flight. Instead of only describing its position as a point in three-dimensional space, we can describe its orientation. As it turns, picture the fly rolling in the same way a plane banks. Its roll is described by the fourth axis, A: the rotational axis around X.

Continuing the plane simile, the fly's pitch is described by the by the fifth axis, B: the rotational axis around Y.

Astute readers will no doubt infer the existence of a sixth axis, C, which rotates about the Z-axis. This is the fly's yaw in our example.

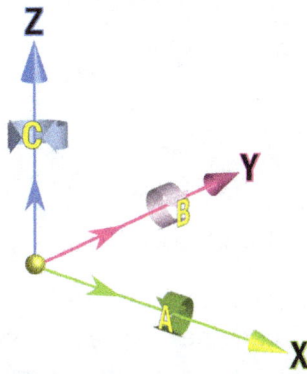

The A, B and C axes are ordered alphabetically to correspond with the X, Y and Z axes. Although there are 6-axis CNC machines, such as Zimmermann's FZ 100 Portal milling machine, 5-axis. configurations are more common, since adding a sixth axis typically offers few additional benefits.

One last note about axis-labeling conventions: in a vertical machining center, the X- and Y-axes reside in the horizontal plane while the Z-axis resides in the vertical plane. In a horizontal machining center, the Z-axis and Y-axis are reversed.

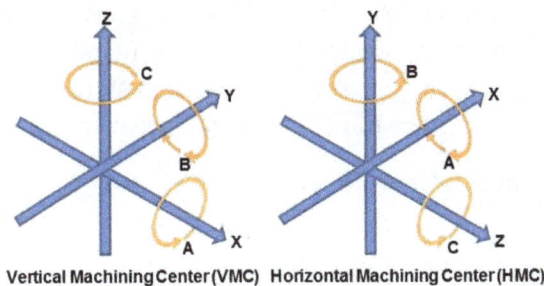

Vertical Machining Center (VMC)   Horizontal Machining Center (HMC)

## 5-Axis Configurations

A 5-axis machine's specific configuration determines which two of the three rotational axes it utilizes.

For example, a trunnion-style machine operates with an A-axis (rotating about the X-axis) and a C-axis (rotating about the Z-axis), whereas a swivel-rotate-style machine operates with a B-axis (rotating about the Y-axis) and a C-axis (rotating about the Z-axis).

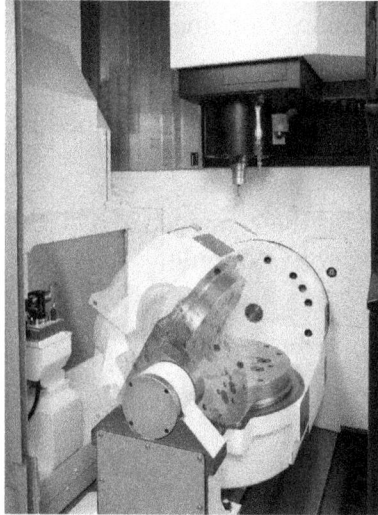

Inside view of the trunnion table of an Okuma MU-4000V 5-axis vertical machining center

The rotary axes in trunnion-style machines are expressed via the movement of the table, whereas swivel-rotate-style machines express their rotary axes by swiveling the spindle. Both styles have their own unique advantages. For instance, trunnion-style machines offer larger work volumes, since there's no need to compensate for the space taken up by the swiveling spindle. On the other hand, swivel-rotate-style machines can support heavier parts, since the table is always horizontal.

## Comparing 3 Axis, 4 Axis and 5 Axis Milling

### 3 Axis

Evolved from the practice of rotary filling, 3 axis machining is an average manual milling technique cutting parts on three axes; the X, Y, and Z axes. Invented in the 1800s, the vertical 3 axis CNC milling machine has come a long way in its capabilities. While many machines have a moving bed, the 3 axis milling centers stand still while the cutter itself operates instead. 3 axis models are multifunctional, accurate machines designed for:

- Automatic/Interactive Operation

- Milling Slots

- Drilling Holes

- Cutting Sharp Edges.

3-axis milling is capable of creating the same products as 4 axis and 5 axis machines, but the 3 axis machines cannot deliver the same level of detail or efficiency as its predecessors. 3 axis can cut individual features during operations, but it cannot match the work of a 5 axis in quality or profitability.

## 4 Axis

Similar to the 3 Axis CNC vertical milling machine, 4 axis micro-milling machines are vertically operated and built to function in a timely fashion. They are user friendly with manual and computer lead operations. Built for detail work and accuracy, 4 axis is a step away from the high quality, precise engravings, drillings, and millings of a 5 axis machine. Despite its short comings, the 4 axis CNC machine can be used for different purposes, including:

- Industry
- Technology Research
- Teaching
- Hobby Prototype Building
- Advertising Design
- Creating Art
- Medical Equipment Creation.

With the ability to operate on wood, foam, composite board, aluminum alloy, and even PCB, 4 axis is a multi-use CNC tool functioning at a slightly higher competency than the 3 axis.

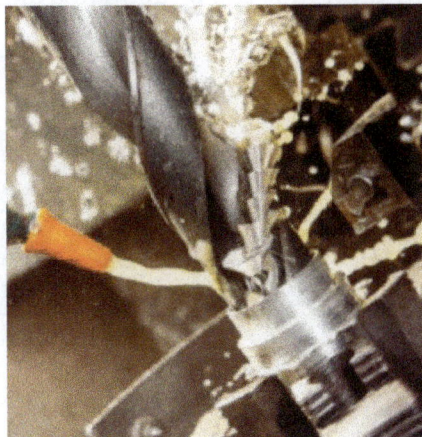

## 5 Axis

The best CNC milling machines available today, 5 axis milling is a fast working, precise, micro machining powerhouse. Owens Industries specializes in 5 axis machining because of its superior functions compared to the 3 axis and 4 axis CNC Machines. A

21st century marvel, a 5 axis CNC machine is capable of impeccable product and parts creation. This extremely fast milling machine, unlike the 3 axis and 4 axis, has the capabilities to efficiently and impressively produce:

- Artificial Bones

- Aerospace Products

- Titanium Creations – For Both Practical and Artistic Purposes

- Oil and Gas Machine Parts

- Car Moulds

- Medical Technology

- Architectural Door Frames

- Military Grade Products.

# Pencil Milling

The pencil milling routine is to finish corners which might otherwise have cusp marks left from previous machining operations. This is ideal for machining into corners where the surface radius is the same as the cutter corner radius. Single pass pencil milling gives a high surface finish ready for polishing. When machining, the tool path maintains climb milling as default and can be used in conjunction with cutter contact angles. As with all tool paths in NCG CAM they can be animated alone or with holders.

## Parallel Pencil Milling

Parallel pencil milling is an extension of pencil milling, in that the user can determine the number and step-over of multiple-passes either side of the pencil tool path. This is particularly useful when the previous cutting tool has not been able to machine all the internal corner radii to size. These multiple passes, will machine the remaining internal radii and any additional material left by the previous cutting tool, machining from the outside into the corner. This creates a good surface finish to the true form and can be used in conjunction with cutter contact angles.

## Corner Offset Machining

Corner offset machining is similar to constant offset machining. However with this technique, rather than starting from an outside boundary and working in towards the Centre of the part, a set of pencil milling passes are created on the features of the part, then a tool path calculated over the whole part from those features. The tool path maintains a constant and equidistant surface finish over the whole part . The resultant surface finish in the corner is significantly better than 3D constant offset machining depending on the shape of the part, as the tool path follows the 3D form and features and can be used in conjunction with cutter contact angles.

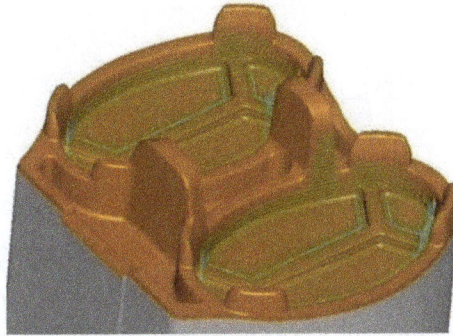

## Boundary Machining

Boundary machining, machines along an open or closed boundary profile. A negative machining thickness can be used to machine at constant depth below the surface being machined and can be used in conjunction with cutter contact angles.

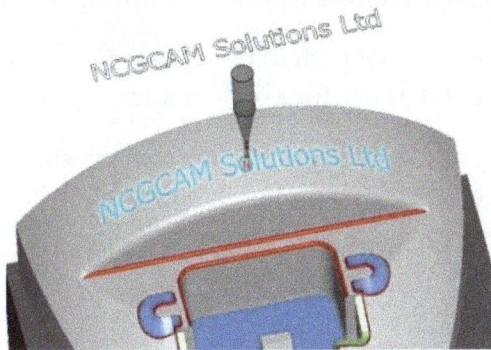

Boundary machining can be used for the machining of mould tool runner detail, or applied to engraving boundary shapes and text which can be generated using the Windows True type TM fonts within the NCG CAM system. The available fonts will depend on the users Windows TM operating system.

## Along Curve Machining and 2D Cutter Compensation

Machining along a curve is just as it says - it is the curve that is machined not the surface data. This will allow a tool path to be generated below the surfaces if needed.

Curves can be read in from the geometry file or extracted from the model. If extracted from the model the curve may be 3D and will be respected as 3D when machined. Curves can also be extracted as 2D curves to be used for 2D machining. These extracted curves contain accurate lines and arcs to get the desired NC Tape file with circular arc moves. There is also a convert curve to boundary function.

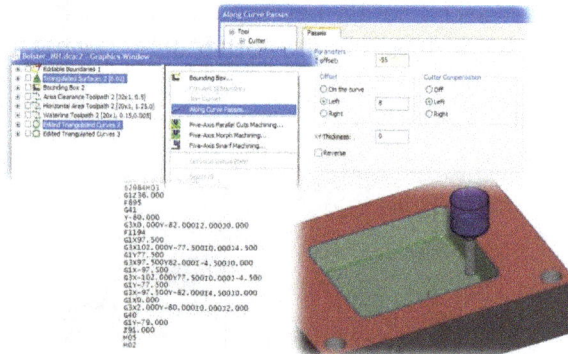

Open curves can be joined to get a continuous profile – often in a model it will be several bits of curve that require joining to reduce the number of retract moves.

The along curve machining supports 2D cutter compensation (G41 & G42 or cutter left/cutter right). This enables 2D profiles to be sized on the machine tool; the tool path has arc fitting for optimised output. Cutter compensation is only available on a 2D curve.

Creating multiple points for start hints allows the user control over the start position and for several curves to be machined within the same operation.

A pass extension will allow the tool path to be extended out (open profiles) to the cutter can be forced to start clear of the part, for a better cutter approach and cutting conditions.

The optional pass overlap allows the cutter to overlap the starting position (closed profiles) to help reduce a 'tool line', giving a better surface finish.

## Part Program

### Types of Part Programming

The part program is a sequence of instructions, which describe the work, which has to be done on a part, in the form required by a computer under the control of a numerical control computer program. It is the task of preparing a program sheet from a drawing sheet. All data is fed into the numerical control system using a standardized format. Programming is where all the machining data are compiled and where the data are translated into a language which can be understood by the control system of the machine tool. The machining data is as follows:

- Machining sequence classification of process, tool start up point, cutting depth, tool path, etc.

- Cutting conditions, spindle speed, feed rate, coolant, etc.

- Selection of cutting tools.

While preparing a part program, need to perform the following steps:

- Determine the startup procedure, which includes the extraction of dimensional data from part drawings and data regarding surface quality requirements on the machined component.

- Select the tool and determine the tool offset.

- Set up the zero position for the work piece.

- Select the speed and rotation of the spindle.

- Set up the tool motions according to the profile required.

- Return the cutting tool to the reference point after completion of work.

- End the program by stopping the spindle and coolant.

The part programming contains the list of coordinate values along the X, Y and Z directions of the entire tool path to finish the component. The program should also contain information, such as feed and speed. Each of the necessary instructions for a particular operation given in the part program is known as an NC word. A group of such NC words constitutes a complete NC instruction, known as block. The commonly used words are N, G, F, S, T, and M. The same is explained later on through examples.

Hence the methods of part programming can be of two types depending upon the two techniques as below:

- Manual part programming.

- Computer aided part programming.

## Manual Part Programming

The programmer first prepares the program manuscript in a standard format. Manuscripts are typed with a device known as flexo writer, which is also used to type the program instructions. After the program is typed, the punched tape is prepared on the flexo writer. Complex shaped components require tedious calculations. This type of programming is carried out for simple machining parts produced on point-to-point machine tool.

To be able to create a part program manually, need the following information:

- Knowledge about various manufacturing processes and machines.
- Sequence of operations to be performed for a given component.
- Knowledge of the selection of cutting parameters.
- Editing the part program according to the design changes.
- Knowledge about the codes and functions used in part programs.

## Computer Aided Part Programming

If the complex-shaped component requires calculations to produce the component are done by the programming software contained in the computer. The programmer communicates with this system through the system language, which is based on words. There are various programming languages developed in the recent past, such as APT (Automatically Programmed Tools), ADAPT, AUTOSPOT, COMPAT-II, 2CL, ROMANCE, SPLIT is used for writing a computer programme, which has English like statements. A translator known as compiler program is used to translate it in a form acceptable to MCU.

The programmer has to do only following things:

- Define the work part geometry.
- Defining the repetition work.
- Specifying the operation sequence.

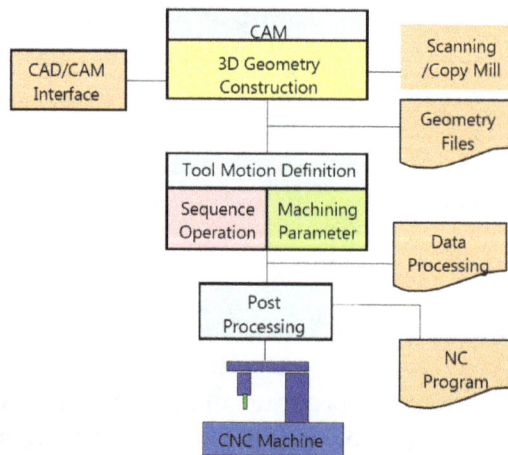

Interactive Graphic System in Computer Aided Part Programming

Over the past years, lot of effort is devoted to automate the part programme generation. With the development of the CAD (Computer Aided Design)/CAM (Computer Aided Manufacturing) system, interactive graphic system is integrated with the NC part programming. Graphic based software using menu driven technique improves the user friendliness. The part programmer can create the geometrical model in the CAM package or directly extract the geometrical model from the CAD/CAM database. Built in tool motion commands can assist the part programmer to calculate the tool paths automatically. The programmer can verify the tool paths through the graphic display using the animation function of the CAM system. It greatly enhances the speed and accuracy in tool path generation.

## Fundamental Elements for Developing Manual Part Programme

The programmer to consider some fundamental elements before the actual programming steps of a part takes place. The elements to be considered are mentioned below.

## Type of Dimensioning System

We determine what type of dimensioning system the machine uses, whether an absolute or incremental dimensional system.

## Axis Designation

The programmer also determines how many axes are availed on machine tool. Whether machine tool has a continuous path and point-to-point control system.

## NC Words

The NC word is a unit of information, such as a dimension or feed rate and so on. A block is a collection of complete group of NC words representing a single NC instruction. An end of block symbol is used to separate the blocks. NC word is where all the machining data are compiled and where the data are translated in to a language, which can be understood, by the control system of the machine tool.

## Block of Information

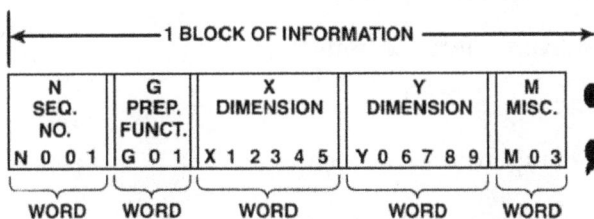

| N SEQ. NO. | G PREP. FUNCT. | X DIMENSION | Y DIMENSION | M MISC. |
|------------|----------------|-------------|-------------|---------|
| N 0 0 1 | G 0 1 | X 1 2 3 4 5 | Y 0 6 7 8 9 | M 0 3 |
| WORD | WORD | WORD | WORD | WORD |

A block of Information: N001 – represents the sequence number of the operation; G01 – represents linear interpolation; X12345 – will move the table in a positive direction along the X-axis; Y06789 – will move the table along the Y-axis; M03 – Spindle on CW and

NC information is generally programmed in blocks of words. Each word conforms to the EIA standards and they are written on a horizontal line. If five complete words are not included in each block.

## Standard G and M Codes

The most common codes used when programming NC machines tools are G-codes (preparatory functions), and M codes (miscellaneous functions). Other codes such as F, S, D, and T are used for machine functions such as feed, speed, cutter diameter offset, tool number, etc. G-codes are sometimes called cycle codes because they refer to some action occurring on the X, Y, and/or Z-axis of a machine tool. The G-codes are grouped into categories such as Group 01, containing codes G00, G01, G02, G03, which cause some movement of the machine table or head. Group 03 includes either absolute or incremental programming. A G00 code rapidly positions the cutting tool while it is above the work piece from one point to another point on a job. During the rapid traverse movement, either the X or Y-axis can be moved individually or both axes can be moved at the same time. The rate of rapid travel varies from machine to machine.

The total numbers of these codes are 100, out of which some of important codes are given as under with their functions:

## G-Codes (Preparatory Functions)

| Code | Function |
|------|----------|
| G00 | Rapid positioning |
| G01 | Linear interpolation |
| G02 | Circular interpolation clockwise (CW) |
| G03 | Circular interpolation counterclockwise (CCW) |
| G20 | Inch input (in.) |
| G21 | Metric input (mm) |
| G24 | Radius programming |
| G28 | Return to reference point |
| G29 | Return from reference point |
| G32 | Thread cutting |
| G40 | Cutter compensation cancel |
| G41 | Cutter compensation left |
| G42 | Cutter compensation right |

G43    Tool length compensation positive (+) direction

G44    Tool length compensation minus (-) direction

G49    Tool length compensation cancels

G53    Zero offset or M/c reference

G54    Settable zero offset

G84    canned turn cycle

G90    Absolute programming

G91    Incremental programming.

## M-Codes (Miscellaneous Functions)

M or miscellaneous codes are used to either turn ON or OFF different functions, which control certain machine tool operations. M-codes are not grouped into categories, although several codes may control the same type of operations such as M03, M04, and M05, which control the machine tool spindle. Some of important codes are given as under with their functions:

| Code | Function |
|------|----------|
| M00 | Program stop |
| M02 | End of program |
| M03 | Spindle start (forward CW) |
| M04 | Spindle start (reverse CCW) |
| M05 | Spindle stop |
| M06 | Tool change |
| M08 | Coolant on |
| M09 | Coolant off |
| M10 | Chuck - clamping M11Chuck - unclamping |
| M12 | Tailstock spindle out |
| M13 | Tailstock spindle in |
| M17 | Tool post rotation normal |
| M18 | Tool post rotation reverse |
| M30 | End of tape and rewind or main program end M98 Transfer to subprogram |
| M99 | End of subprogram. |

## Tape Programming Format

Both EIA and ISO use three types of formats for compiling of NC data into suitable blocks of information with slight difference.

## Word Address Format

This type of tape format uses alphabets called address, identifying the function of numerical data followed. This format is used by most of the NC machines, also called variable block format. A typical instruction block will be as below:

> N20 G00 X1.200 Y.100 F325 S1000 T03 M09 <EOB>

> or

> N20 G00 X1.200 Y.100 F325 S1000 T03 M09;

The MCU uses this alphabet for addressing a memory location in it.

## Tab Sequential Format

Here the alphabets are replaced by a Tab code, which is inserted between two words. The MCU reads the first Tab and stores the data in the first location then the second word is recognized by reading the record Tab. A typical Tab sequential instruction block will be as below :

> >20 >00 >1.200 >.100 >325 >1000 >03 >09

## Fixed Block Format

In fixed block format no letter address of Tab code are used and none of words can be omitted. The main advantage of this format is that the whole instruction block can be read at the same instant, instead of reading character by character. This format can only be used for positioning work only. A typical fixed block instruction block will be as below:

> 20 00 1.200 .100 325 1000 03 09 <EOB>

## Machine Tool Zero Point Setting

The machine zero point can be set by two methods by the operator, manually by a programmed absolute zero shift, or by work coordinates, to suit the holding fixture or the part to be machined.

## Manual Setting

The operator can use the MCU controls to locate the spindle over the desired part zero and then set the X and Y coordinate registers on the console to zero.

## Absolute Zero Shift

The absolute zero shift can change the position of the coordinate system by a command in the CNC program. The programmer first sends the machine spindle to home zero position by a command in the program. Then another command tells the MCU how far from the home zero location, the coordinate system origin is to be positioned.

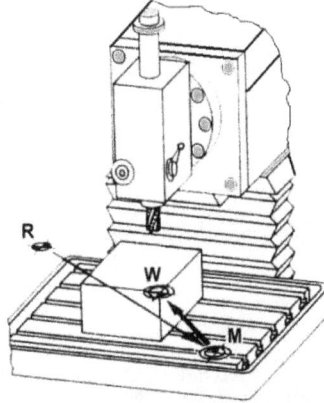

Machine Tool Zero Point Setting

R = Reference point (maximum travel of machine)

W = Part zero point workpice coordinate system

M = Machine zero point (X0, Y0, Z0) of machine coordinate system

The sample commands may be as follows:

N1 G28 X0 Y0 Z0 (sends spindle to home zero position or Return to reference point).

N2 G92 X3.000 Y4.000 Z5.000 (the position the machine will reference as part zero or Programmed zero shift).

## Coordinate Word

A coordinate word specifies the target point of the tool movement or the distance to be moved. The word is composed of the address of the axis to be moved and the value and direction of the movement.

Example,

X150 Y-250 represents the movement to (150, – 250). Whether the dimensions are absolute or incremental will have to be defined previously using G-Codes.

## Parameter for Circular Interpolation

These parameters specify the distance measured from the start point of the arc to the center. Numerals following I, J and K are the X, Y and Z components of the distance respectively.

## Spindle Function

The spindle speed is commanded under an S address and is always in revolution per minute. It can be calculated by the following formula:

$$\text{Spindle Speed} = \frac{\text{Surface cutting speed in } m/min \times 1000}{\pi \times \text{Cutter Diameter in mm}}$$

Example,

S1000 represents a spindle speed of 1000 rpm.

## Feed Function

The feed is programmed under an F address except for rapid traverse. The unit may be in mm per minute or in mm per revolution. The unit of the federate has to be defined at the beginning of the programme. The feed rate can be calculated by the following formula:

$$\text{Feet Rate} = \frac{\text{Chip Load}}{\text{Tooth} \times \text{No .of tooth} \times \text{Spindle speed}}$$

Example

F100 represents a feed rate of 100 mm/min.

## Tool Function

The selection of tool is commanded under a T address.

### Work Settings and Offsets

All NC machine tools require some form of work setting, tool setting, and offsets to place the cutter and work in the proper relationship. Compensation allows the programmer to make adjustments for unexpected tooling and setup conditions. A retraction point in the Z-axis to which the end of the cutter retracts above the work surface to allow safe table movement in the X-Y axes. It is often called the rapid-traverse distance, retract or work plane. Some manufacturers build a work piece height distance into the MCU (machine control unit) and whenever the feed motion in the Z-axis will automatically be added to the depth programmed.

When setting up cutting tools, the operator generally places a tool on top of the highest surface of the work piece. Each tool is lowered until it just touches the work piece surface and then its length is recorded on the tool list. Once the work piece has been set, it is not generally necessary to add any future depth dimensions since most MCU do this automatically.

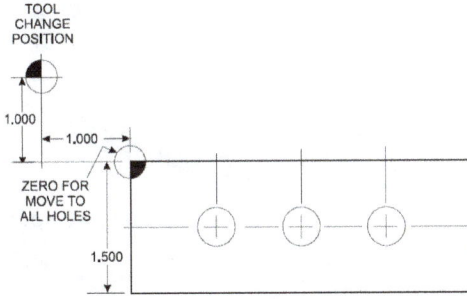

Work Settings

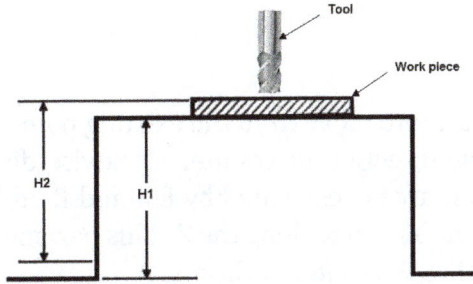

Offsets

## Rapid Positioning

This is to command the cutter to move from the existing point to the target point at the fastest speed of the machine.

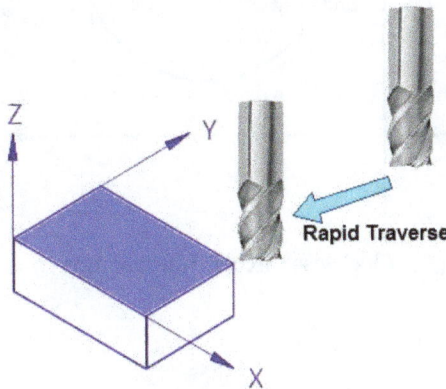

Rapid Positioning

## Linear Interpolation

This is to command the cutter to move from the existing point to the target point along a straight line at the speed designated by the F address.

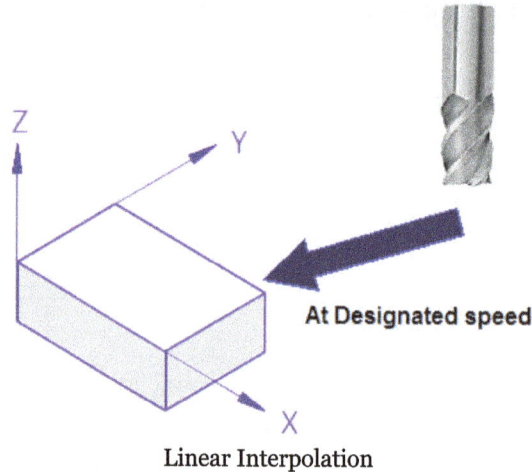

Linear Interpolation

## Circular Interpolation

This is to command the cutter to move from the existing point to the target point along a circular arc in clockwise direction or counter clockwise direction. The parameters of the center of the circular arc is designated by I, J and K addresses. I is the distance along the X-axis, J along the Y, and K along the Z. This parameter is defined as the vector from the starting point to the center of the arc.

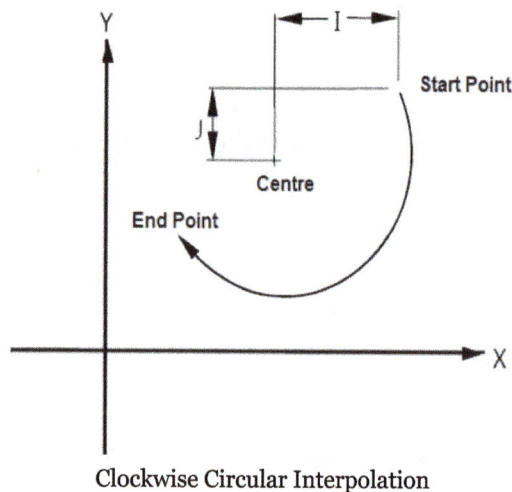

Clockwise Circular Interpolation

## Circular Interpolation

In NC machining, if the cutter axis is moving along the programmed path, the dimension of the work piece obtained will be incorrect since the diameter of the cutter has not be taken in to account. What the system requires are the programmed path, the cutter diameter and the position of the cutter with reference to the contour. The cutter diameter is not included in the programme. It has to be input to the NC system in the tool setting process.

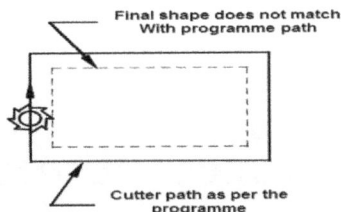

Figure 4.9 : Tool Path without Cutter Compensation

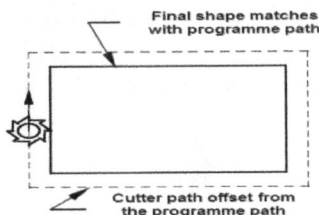

Tool Path without Cutter Compensation

# References

- Kramer, Thomas R. (1992). "Pocket Milling with Tool Engagement Detection". Journal of Manufacturing Systems. 11 (2): 112–123. doi:10.1016/0278-6125(92)90042-E. Retrieved 11 March 2015

- Everything-you-need-to-know-about-cnc-machines: creativemechanisms.com, Retrieved 18 July 2018

- What-is-cnc-machining: astromachineworks.com, Retrieved 22 May 2018

- Held, Martin (1991). "A geometry-based investigation of the tool path generation for zigzag pocket machining". The Visual Computer. 7 (5-6): 296–308. doi:10.1007/BF01905694. Retrieved 11 March 2015

- What-is-cutter-location-cl-data-file: qhunt.com, Retrieved 18 July 2018

- Running-programs-with-direct-numerical-control: mmsonline.com, Retrieved 22 April 2018

- Ncg-cam-pencil-milling-and-other-corner-sequence-machining: inas.ro, Retrieved 08 July 2018

- Bieterman, Michael B.; Sandstrom, Donald R. (Nov 11, 2003). "A Curvilinear Tool-Path Method for Pocket Machining". Journal of Manufacturing Science and Engineering. 125: 709–715. doi:10.1115/1.1596579

- Inventory-management-101-master-production-schedule-mps-explained: optiproerp.com, Retrieved 28 May 2018

- Milling-machine-definition-process-types: engineeringarticles.org, Retrieved 18 May 2018

# 6

# Enterprise Resource Planning

Enterprise resource planning (ERP) refers to the integrated management of business processes, achieved by a combination of software and technology. All the diverse aspects of enterprise resource planning have been carefully analyzed in this chapter, such as enterprise planning systems, ERP modeling, manufacturing resource planning, master data management, etc.

## Enterprise Planning Systems

Enterprise Resource Planning (ERP) systems are computer software programs designed to integrate the multiple individual functional requirements, and operational information needs of a business. The primary purpose of an ERP system is to establish a central data resource for all the information that is required to be recorded for day to day operations, and to provide that information to the functional elements of the organization as needed to fulfill the goals of the business. They are typically implemented using relational database schema, which allow for the utilization of a central data set in all business process transactions and the retention of all critical business information. One principle advantage, over a business system built up from individual software's, is that they can provide reports which provide an accurate and real or quasi-real time view of the company and its operational condition. Managers can review reports from a single database, which reflects real-time or very near real time information, rather than the frequently contradictory condition assessments derived from several independent data sources.

Early implementers of ERP systems were primarily large business enterprises, and accordingly, the marketplace for appropriate systems was limited to a few large scale software systems such as BAAN, SAP and J.D. Powers .The bulk of the decision process to adopt such a system was therefore based in economic decision making and identification of the most closely related system in terms of broad based business culture issues. As the supply chain deepened to include strategic partnerships with small and medium-sized enterprises, there was a growing expectation that these small and medium-sized businesses should adopt ERP systems which emulated, and integrated with, the information requirements of the large business partner. This impetus, and that driven by the desire of small businesses to have better and more timely understanding of the status of production, logistics, finances led to a high demand for ERP systems that addressed the unique needs of these businesses in a cost effective manner.

The proliferation of new offerings of Enterprise Resource Planning systems; implemented as software applications, or more recently as ASP (Application Service Provider) hosted software, has added an additional level complexity to the selection and best-value decision making process identify the right ERP business software for small businesses. The current market offers in excess of 200 individual software systems, ranging from large scale systems with multi-faceted capability sets, typically aimed at large business enterprises, to software systems aimed at the small business consumer. All businesses and in particular small businesses are faced with this daunting selection of offerings, and consequentially, the selection process is a major challenge to organizations which are about to embark on the implementation of ERP in order to improve their integrated business systems. This selection process is a major challenge to any business the outcome of which may realize either huge potential benefits or create great risks for the enterprise. As early as 1998 (Martin, 1998) reported the tardiness of implementation and cost overrun liability at 90%, and later papers reveal that 70% of implementations "fail to deliver anticipated benefits" (Al-Mashari & Zairi, 2000). Overall, anecdotal evidence suggests that, even today after years of fit analysis improvements in the ERP selection process, up to 60% of all implementations are regarded as failures in varying degrees.

The global budget for ERP implementation is projected to exceed $36 Billion in the next decade. The worldwide market for enterprise resource planning (ERP) systems grew at a 4.8% compound annual rate, rising from $16.7 billion in 2005 to more than $21 billion in 2010, according to a study. A methodology which reliably identifies an ERP system for a small or medium-sized business enterprise and reduces risk in implementation, will serve to provide practical benefits to industry.

An illustrative definition for ERP is: "An enterprise-wide set of management tools that balances demand and supply, containing the ability to link customers and suppliers into a complete supply chain, employing proven business processes for decision making, and providing high degrees of cross-functional integration among sales, marketing, manufacturing, operations, logistics, purchasing, finance, new product development, and human resources, thereby enabling people to run their business with high levels of customer service and productivity, and simultaneously lower costs and inventories; and providing the foundation for effective ecommerce."

Historically, ERP systems can trace their origins from the development of management and resource utilization theories in the post Second World War period. As new concepts, such as Total Quality Management (TQM) and Statistical Process Control (SPC) were developed and widely introduced, it became evident that the data derived from these new techniques had the potential to provide extended information about the efficiency of the enterprise as a whole and its business posture, well beyond the direct effects of improved product quality. Beginning in the late 1940's W. Edwards Deming encouraged the use of SPC theories and production control techniques in the US; many of which had been proposed by a fellow physicist/statistician Walter A. Shewhart, in the late 1930's. The famous; Plan, Do, Check, Act cycle which Deming advocated, was initially formulated by Shewhart in their joint work on the subject of scientific inference (Shewhart, 1939). Shewhart's work demonstrates an early understanding of the key influence that data have with respect to the success of an organization, and proposes two essential functions/attributes that are addressed by ERP systems:

1. "Data have no meaning apart from their context."

2. "Data contain both signal and noise. To be able to extract information, one must separate the signal from the noise within the data."

Deming was an advocate for SPC based management protocols which helped increase the efficiency of American War industry in the 1940's, and he studied and applied the theories of common and special variation. Deming spent much of the WW II involved in the education and training of industrial managers and during this time began to identify some of the key concepts of his fourteen points, most particularly, the critical role of upper management's essential 'buy-in' when resorting to large scale changes in business philosophy. There is a widely held consensus that management "buy-in" and its cultural effect on a company is a key factor in the success or failure of ERP implementations.

It has been suggested that, 'management programs', are subject to a life cycle simply stated as; inception, growth, maturity and decline (Crandall, 2005). However, despite the fears to the contrary, some management systems do have extended utility, and indeed actually evolve into better and more widely applicable systems. The MRP – MRP II – ERP development path represents just such an evolutionary process. The key factor differentiating MRP II from ERP systems are in fact platform dependent limitations – had the computing environment been more advanced or consolidated around a common architecture in 1980, MRP II could have remained the dominant system for all enterprise system developments. The 'genetic' similarities of MRP II and ERP are apparent in their common approach to addressing system needs, in both cases the systems are predominantly modular in nature, and the systems rely upon the relational database management system (RDBMS) approach to storing and sharing data. MRP is concerned primarily with manufacturing materials while MRPII is concerned with the coordination of all manufacturing activities, including logistics, finance, and human resources. The goal of MRPII, and of ERP in a broader sense, is to provide consistent data to all decision makers in the manufacturing process as the product is manufactured.

MRP and MRP II are still in current use either as standalone solutions, or as subcomponents of a more complex suite of management tools which can be described in their entirety as an Enterprise Resource Planning software system. Despite the recognition of the weaknesses of earlier MRP and MRP II systems, and the development of solutions to these weaknesses, some systems have in fact enjoyed prolonged utility. Heritage MRP systems are frequently maintained because they are the 'optimal' solution for a particular enterprise and remain appropriate due to lower levels of need which are consistent with the maturity level of the enterprise.

The wholesale development of business information systems, enterprise systems and ERP systems has led to a highly competitive marketplace. Adopting an enterprise software system introduces a number of cost factors to an organization; cost of acquisition and maintenance, cost of implementation, cost of operational disruption, costs of morale/psychological impacts, and for a small business can introduce the risk of failure of the entire business.

## ERP System Selection Methodology

The acquisition of an ERP package, the decision to deploy one already owned into a different division/plant of a firm or, as in Product Life Cycle, the need to upgrade function, form or fitness of an existing package in use, is the typical initial phase of these projects' life cycle.

This phase is a project in itself, it is often assumed to be the first phase in the overall implementation and its main deliverable is the selection of the "best" ERP package to be implemented.

But what is the best package?

Most ERP technology embodies a proven set of Logistics, Manufacturing, Customer Relations and Financial practices, therefore when you embark in an ERP system purchase you are implicitly and explicitly accepting those practices as the business process framework for your firm. The packages, of course, are not all alike; the set of available proven practices varies according to the size (scale) of the firm and most packages offered less functionalities for small firm (less than 50 employees) versus a full set of functionalities for a larger one (more than 500 employees) and the cost of ownership varies accordingly. Furthermore there are packages that focus on a particular industry sector (textiles, automotive, electronics, etc.) and hence the practices they support are very specific to the sector in question and may not be universally applicable to others.

If there is a scale or industry focus mismatch (although it does not happens that often) your implementation project will suffer and you are almost guaranteed to be over budget by trying to retrofit the package and eliminate the mismatch with functionality's modifications and, what is worse, still deploy a tool that does not serve the needs of the business.

The problem is that, in most cases, those stakeholders who have responsibility and authority in selecting the "best" package are not the ones with responsibility and authority in using it. Most methodologies correctly identify the former group but tend to ignore the latter one.

As we all know, the influence of stakeholder is high at the project initiation as it is low the cost of changes; unfortunately the "using" stakeholders come to exert their influence in the intermediate phases of the project life cycle (such as in Conference Room

Pilot mode), when satisfying it is significantly more costly; hence the implementation project is guaranteed to run into problem unless this fundamental issue is addressed.

Some of the aspects that companies should consider before deciding on an ERP system are as follows.

- Involving key staff from critical departments in the project.

- Having adequate discussions to have a clear understanding of all essential functionalities.

- Getting expert advice from vendors but also involve the in-house personnel validate the same for suitability.

- Estimating the Total Cost of Ownership (TCO) well in advance to avoid overheads.

- Validating suggestions from all the members involved in the project.

- Understand the present IT systems and evaluate the additional requirements for the ERP implementation.

- Setting the expectations right, before the project commencement. This ensures that the teams across the organization plan accordingly.

After evaluating the aforementioned aspects, the next step is to identify potential vendors and select the most suitable solution. The following is a list of activities a company should undertake, before arriving at a solution:

- Perform an exhaustive analysis of functional and technical requirements.

- Drafting a detailed Request For Proposal (RFP) that states without any ambiguity, all the requirements while taking into consideration various factors like cost, technology, ROI etc.

- Perform a detailed market research to get a list of all potential vendors who can serve the purpose and circulating the RFP to all the selected vendors.

- View a demonstration of the ERP solution and discuss with the vendor in detail about tailoring the solution to the requirements.

- Choosing a vendor who can provide the most effective solution that can meet the maximum requirements.

- Calculate the total cost of ownership (TCO) and negotiating the price and contract. A legal advisory firm can help cross check and include all relevant clauses required.

- Devising effective plans and implementation strategies ensures flawless execution.

# ERP Modeling

The generic ERP package represents the commonly operated business model of the organization. It is built with the function models like the Finance, Materials, Marketing, Sales and Personnel and their sub-modules. These modules are then integrated to perform, ensuring data and information consistency and concurrency.

The seamless integration of the modules allows the user at any level to take a micro and a macro view of the function and process view of the transaction across the function. A typical ERP solution has the following modules:

- Business forecasting, planning and control (Business)

- Sales, distribution, invoicing (Sales)

- Production planning and control (Production)

- Materials management (Materials)

- Finance and accounting (Finance)

- Personnel management (Personnel)

ERP delivery models have evolved over the years and given the organizations many options to choose from. The basic points of consideration include work overload, cost economy, flexibility of operation and data security. Before you choose any model, you must consider the implication it has on your employees, suppliers and customers.

Implementation procedure for every model will vary depending upon the complexity and risk factor involved. The training requirement will also be different for different models. Work out the cost, long-term strategy and plan the implementation process.

Various delivery models are discussed below.

## On-premise

On-premise ERP software means that the company purchases the software and installs it in its own servers. The company has to further invest in ERP tools, add-ons, long-term and short-term server maintenance. On-premise software gives the leverage of customizing the software as per the typical needs of an organization. Though, it may look complex to manage, many businesses feel that it scores an edge over other delivery models.

Advantages

- Your business data is secured within your company-owned servers.

- Your data is secured as you don't have to worry about data security threats and data leaks.

- You save significant cost as you don't have to invest in data security software and manpower to handle the same.

Disadvantages

- You have to spend significant time in implementation, setting up the system and loading new modules.

- Your infrastructure cost is high as you have to invest in servers to increase data handling capacity.

- You have to allocate extra IT staff to run the software on your system.

## Hosted

All ERP software needs a server to run. In hosted ERP model, you own the software but not the server. You rather use third party servers by paying a monthly fee. All your data is located in these third-party secured servers. Data back-up, data security and server maintenance is taken care by the vendor itself. In hosted services, software is accessed through virtual private network rather than through internet. It is gradually being replaced web-based cloud-hosted ERP models.

Advantages

- Vendors make sure that your data is secured, with a balanced architecture and an up-time guarantee.

- Vendors ensure that your software is easily accessible to your employees.

- Vendors have sufficient experience in the industry to keep your software up and running.

- Your in-house team is relieved of maintenance.

Disadvantages

- Your data is at risk if vendors server is hacked.

- If there is any technical issue with the vendor's servers, your business functions will come to a halt.

## Public Cloud

Public cloud ERP software is owned and remotely hosted by the vendor's servers. It is accessed over internet on a subscription basis. The subscription fee is inclusive of access to hardware, software and other resources essential for running the software. This model is commonly known as SaaS i.e. Software-as-a-Service. This technology allows the applications to be installed across multiple servers and data base resources. It is especially useful when the business is spread across multiple locations.

Advantages

- You have 24*7 access to cloud services.

- You don't have to worry about hardware, software and upgrades thereby reducing your infrastructure investment for setting up the ERP system.

- You save time and money on installation and maintenance.

- Cloud services can be delivered through any device connected with web without using complicated VPN network.

- Real-time data access.

- You have the flexibility to change your service provider.

- You only pay for what you use.

Disadvantage

- Cloud host providers offer their services to multiple users at a time which help them to keep their cost low at the expense of limited flexibility. It becomes a cause of concern for organizations who need customization to adjust their business needs.

## Private Cloud

Private cloud based software allows the organization to host the application on a private server located on-premise. The ownership of the software lies with the vendor only. In other cases, the private server can be located at a remote location under the control of the vendor. Here, the vendor remotely manages and updates the server meant for you. It offers the flexibility of enabling customizations and developing applications without compromising on other facilities offered by SaaS solution.

Advantages

- As the cloud server is private, you hold a higher degree of control over it.

- Your data is backed by more security.

- All other advantages are in common with public cloud.

Disadvantages

- Unlike the case in public clouds, private clouds are costly to maintain,

- Capacity addition becomes a problem for private cloud owners as it needs more investment in servers and maintenance. Moreover, servers are under-utilized most of the time.

## Hybrid

Hybrid ERP solution is a combination of on-premise and cloud format. It allows customers to migrate from one delivery model to another without losing data or functionality. The associated cloud can be private or public depending upon the needs of the organization.

Advantages

- When the saturation limit for existing hardware set-up is reached in an on-premise delivery model, companies look for better ROI for further investment. Hybrid set-up allows the organization to migrate non-critical processes to the cloud while retaining the critical processes on-premise. Here, you can pay a meager fee to deploy certain processes on the cloud i.e. you pay for what you use.

- Testing is usually required during ERP implementation, new releases and upgrades. On-premise set-up makes it difficult for companies to test the processes at full-load due to limited storage capacity. Hybrid set-up allows the companies to commission a testing environment on the cloud which needs a huge capacity on a temporary basis. After the testing is over, the cloud services can be decommissioned. It helps the companies to save significant expenditure in permanent capacity expansion.

Hybrid ERP is the best delivery model for a mid-size to big organization as it follows a balanced approach. It does not specifically have any disadvantage except the similar challenges faced by on-premise and cloud systems.

## Enterprise Resource Planning

ERP software is comprised of powerful and strategic business process management tools that can be used to manage information within an organization. While every company and organization operating today is different, they all face a common challenge: in order to stay competitive in today's business environment, they need a dependable and efficient way to store and access information. That's where ERP systems come into play. ERP systems integrate all facets of an enterprise into one comprehensive information system that can be accessed by individuals across an entire organization.

With effective ERP software in place, business owners and leaders can automate and streamline tedious back office tasks, help employees become more productive and successful in their roles, and get real-time visibility into the inner workings of their operations.

## Evolution of ERP & Current Trends

The term "ERP" or "Enterprise Resource Planning" was originally coined by industry analyst, The Gartner Group, in the 1990s. It evolved from MRP, a term that was already well-known in business at that time. MRP stands for both Material Requirements Planning (MRP) and Manufacturing Resource Planning (MRPII). These systems were created back in the 1960s when manufacturing-based companies were looking for ways to improve efficiency and decision-making for production line managers.

In the 1990s, The Gartner Group and other businesses sought to apply MRP systems to other business types, and desired to expand capabilities and processes to other areas within an organization, and thus ERP as we know it today was born. In its early days of existence, ERP focused on organizing data and streamlining processes that related to back-office areas, such as inventory management, fulfillment, purchasing, human resources, accounting, IT, manufacturing, planning and scheduling, and other related areas.

Later, with the introduction and widespread use of the Internet, ERP was expanded further to include other areas of a business, such as customer relationship management (CRM), supplier relationship management (SRM) and supply chain management (SCM), human capital management (HCM), business intelligence and ecommerce.

Today, ERP systems integrate into all areas and functions within an organization, with the primary purpose being to help leaders and managers better understand all moving parts of their operations, identify opportunities, and make more informed decisions that will ultimately have an impact on the future success and viability of their businesses.

Despite the use of the word enterprise in the name, ERP systems are used by businesses of all sizes, large and small. There are two primary types of ERP systems being implemented at organizations today, On-premises and Cloud-based.

"The Compass software reports have been especially eyeopening. For example, the low margin sales and cost change reports have helped us stay on top of pricing," said Braaten. "We've increased our margins by adjusting items that weren't priced correctly. Our soft lines margins have increased by 6 percent companywide."

"Additionally, in one of our locations we saw an $80,000 increase in margin alone with the implementation of Epicor solutions, all because we have reliable information to see where to make modifications within the business," said Braaten.

An ERP system is made up of applications and tools that help all areas of your business communicate with each other more effectively. ERP systems integrate all facets of an enterprise into one comprehensive information system. Employees in planning and scheduling, for example, have access to the same data as the staff in financial management for their specific needs. All data is available in real-time, which enables employees to make faster, more informed business decisions. With ERP systems, all vital business functions—estimating, production, finance, human resources, marketing, sales, purchasing—share a central source of up-to-the-minute information. Enterprise resource planning systems streamline the collection, storage and use of your organization's data. The right ERP system can help you collect and store data into one centralized place from areas such as:

- Finance & Accounting

- Human Resources

- Customer Relationship Management

- Production Management

- Business Intelligence

- Warehouse Management

- Inventory Management

- Supply Chain Management

- Point-of-Sale (POS)

- E-Commerce

## Primary Benefits of ERP Systems

Why are more businesses of all sizes implementing ERP systems today than ever before? Here are some of the main reasons and benefits why organizations use ERP systems:

- Provide business leaders with real-time visibility into their operations.

- Provide business leaders and teams with instant access to their global supply chains.

- Enable business leaders to identify challenges, uncover opportunities, and make faster decisions that impact different areas of their businesses.

- Help automate and streamline tedious tasks and redundant processes.

- Give employees the tools and data they need to be successful.

- Provide a single point of truth for organizations.

- Can often be accessed from anywhere (off-site and from mobile devices).

- Help increase productivity among your team.

- Make it easier for teams to collaborate with each other, and with third-party vendors.

- Offer powerful reporting and forecasting tools that you can use to make informed decisions about the future of your business.

- Keep data secure, and help you ensure that your business continues to operate in compliance with global regulatory laws and guidelines.

## Deciding If/when your Business Needs ERP

Every business is unique and faces different challenges at different times, so the question is, how do you decide if and when investing in Enterprise Resource Planning is right for your business.

If you're able to check off most of the items on this list, it's probably safe to start evaluating ERP software providers and working to allocate the resources needed for deployment:

- Your team members are spending too much time on tasks you know could be automated and streamlined.

- You don't have easy access into the data you need to make informed decisions about your business.

- You work with vendors and third-party applications across the globe.

- You have a lot of different software tools and processes that you've adopted and implemented for your business over the years, but they are not connected to each other.

- You don't know what your inventory levels really look like on a daily basis.

- You're personally spending too much time searching for information, trying to boost productivity and efficiencies, and integrating new tools that are needed in order to scale. Your teams can't easily collaborate or share information with each other.

- You can't access essential business data and information when you're off-site.

- You're having trouble keeping up with changes in regulatory compliance.

- You're finding or addressing problems after it's too late; in other words, you're not able to be as proactive as you'd like when it comes to identifying problems that need to be fixed in order to keep your operations running smoothly.

If you can check off at least sixty or seventy percent of the list above, it's time to start exploring vendors.

In order to justify the investment needed in order to use an ERP system at your business, use the checklist provided above. List the specific challenges you think your organization is having, and talk with your decision-maker about how you think an ERP solution could help.

## Types of Deployment Options for ERP Solutions

There are two primary types of deployments for ERP solutions available to businesses today: Cloud-Based and On-Premises.

On-Premises ERP is deployed locally on your hardware and servers, and managed by your IT staff. Businesses that choose this option want greater autonomy over their implementation.

Cloud ERP software looks and works the same as traditional ERP. The only difference is how it is deployed. With cloud ERP, instead of hosting your servers and storage hardware on-site, your ERP provider hosts this for you. To access your ERP system, you simply sign in to a website hosted online (in the cloud).

Cloud-based ERP brings enterprise-grade security to protect today's businesses, as well as a lower cost of ownership, ease of use, and configuration flexibility. It also provides you with real-time access and visibility into your business information via your cell phone or other mobile device, no matter where you are in the world.

## Choosing the Right Solution

If you're looking for your first ERP solution or looking to upgrade from an existing system, the evaluation, selection, and implementation process is a long-term strategic decision for your organization.

To help you through this process, here are eight simple steps for a successful ERP system selection:

- Step 1: Evaluation- Get members of your leadership together and get the initial conversations going about ERP. Make sure everyone has transparency into how you are evaluating providers.

- Step 2: Make An Assessment- Look critically at your business and operations to determine what's working well, what challenges you're experiencing, and what you think you need in order to scale or make improvements.

- Step 3: Establish Criteria- Develop a standard criteria to use when evaluating vendors. Criteria can include features, price, platform, and anything else your team thinks is important when making evaluations.

- Step 4: Schedule Consultations- Schedule phone calls, in-person meetings, or online demos with sales representatives and product experts at the vendors you're considering. Get a first impression about each company and try to make note of the differences between the vendors you talk to.

- Step 5: Create a Shortlist- Narrow your list down to 2-3 candidates that you'd like to follow-up with.

- Step 6: Make Contact - Contact vendors on your shortlist and schedule follow-up calls to learn more about how their products could help your business.

- Step 7: Prepare Questions That Address Your Concerns- Prepare specific questions that you'd like each vendor to address during the follow-up call.

- Step 8: Check Vendor References- Talk to other companies that have implemented ERP systems from the vendors you're considering. Find out what they like, what they don't like, what they weren't expecting, if they'd actually recommend the vendor you're asking about, etc. Get as much information from them as you can in order to make a final decision. Once you've selected the ERP vendor for your business and needs, spend the time to scope the entire project, from start to finish. Work with your ERP vendor to map your critical milestones and chart your course to success.

## Material Requirements Planning

Material requirements planning (MRP) is a computer-based inventory management system designed to assist production managers in scheduling and placing orders for items of dependent demand. Dependent demand items are components of finished goods—such as raw materials, component parts, and subassemblies—for which the amount of inventory needed depends on the level of production of the final product. For example, in a plant that manufactured bicycles, dependent demand inventory items might include aluminum, tires, seats, and bike chains.

The first MRP systems of inventory management evolved in the 1940s and 1950s. They used mainframe computers to explode information from a bill of materials for a certain finished product into a production and purchasing plan for components. Before long, MRP was expanded to include information feedback loops so that production personnel could change and update the inputs into the system as needed. The next generation of MRP, known as manufacturing resources planning or MRP II, also incorporated marketing, finance, accounting, engineering, and human resources aspects into the planning process. A related concept that expands on MRP is enterprise resources planning (ERP), which uses computer technology to link the various functional areas across an entire business enterprise.

MRP works backward from a production plan for finished goods to develop requirements for components and raw materials. MRP begins with a schedule for finished
goods that is converted into a schedule of requirements for the subassemblies, the component parts, and the raw materials needed to produce the final product within the
established schedule. MRP is designed to answer three questions: what is needed? how
much is needed? and when is it needed?.

MRP breaks down inventory requirements into planning periods so that production can be
completed in a timely manner while inventory levels—and related carrying costs—are kept
to a minimum. Implemented and used properly, it can help production managers plan for
capacity needs and allocate production time. But MRP systems can be time consuming and
costly to implement, which may put them out of range for some small businesses. In addition, the information that comes out of an MRP system is only as good as the information
that goes into it. Companies must maintain current and accurate bills of materials, part
numbers, and inventory records if they are to realize the potential benefits of MRP.

## MRP Inputs

The information input into MRP systems comes from three main sources: a bill of materials, a master schedule, and an inventory records file. The bill of materials is a listing
of all the raw materials, component parts, subassemblies, and assemblies required to
produce one unit of a specific finished product. Each different product made by a given
manufacturer will have its own separate bill of materials. The bill of materials is arranged in a hierarchy, so that managers can see what materials are needed to complete
each level of production. MRP uses the bill of materials to determine the quantity of
each component that is needed to produce a certain number of finished products. From
this quantity, the system subtracts the quantity of that item already in inventory to determine order requirements.

The master schedule outlines the anticipated production activities of the plant. Developed using both internal forecasts and external orders, it states the quantity of each
product that will be manufactured and the time frame in which they will be needed. The
master schedule separates the planning horizon into time "buckets," which are usually
calendar weeks. The schedule must cover a time frame long enough to produce the
final product. This total production time is equal to the sum of the lead times of all the
related fabrication and assembly operations. It is important to note that master schedules are often generated according to demand and without regard to capacity. An MRP
system cannot tell in advance if a schedule is not feasible, so managers may have to run
several possibilities through the system before they find one that works.

The inventory records file provides an accounting of how much inventory is already on
hand or on order, and thus should be subtracted from the material requirements. The
inventory records file is used to track information on the status of each item by time
period. This includes gross requirements, scheduled receipts, and the expected amount

on hand. It includes other details for each item as well, like the supplier, the lead-time, and the lot size.

## MRP Processing

Using information culled from the bill of materials, master schedule, and inventory records file, an MRP system determines the net requirements for raw materials, component parts, and subassemblies for each period on the planning horizon. MRP processing first determines gross material requirements, then subtracts out the inventory on hand and adds back in the safety stock in order to compute the net requirements.

The main outputs from MRP include three primary reports and three secondary reports. The primary reports consist of: planned order schedules, which outline the quantity and timing of future material orders; order releases, which authorize orders to be made; and changes to planned orders, which might include cancellations or revisions of the quantity or time frame. The secondary reports generated by MRP include: performance control reports, which are used to track problems like missed delivery dates and stock outs in order to evaluate system performance; planning reports, which can be used in forecasting future inventory requirements; and exception reports, which call managers' attention to major problems like late orders or excessive scrap rates.

Although working backward from the production plan for a finished product to determine the requirements for components may seem like a simple process, it can actually be extremely complicated, especially when some raw materials or parts are used in a number of different products. Frequent changes in product design, order quantities, or production schedule also complicate matters. The importance of computer power is evident when one considers the number of materials schedules that must be tracked.

## Benefits and Drawbacks of MRP

MRP systems offer a number of potential benefits to manufacturing firms. Some of the main benefits include helping production managers to minimize inventory levels and the associated carrying costs, track material requirements, determine the most economical lot sizes for orders, compute quantities needed as safety stock, allocate production time among various products, and plan for future capacity needs. The information generated by MRP systems is useful in other areas as well. There is a large range of people in a manufacturing company that may find the use of information provided by an MRP system very helpful. Production planners are obvious users of MRP, as are production managers, who must balance workloads across departments and make decisions about scheduling work. Plant foremen, responsible for issuing work orders and maintaining production schedules, also rely heavily on MRP output. Other users include customer service representatives, who need to be able to provide projected delivery dates, purchasing managers, and inventory managers.

MRP systems also have several potential drawbacks. First, MRP relies upon accurate input information. If a small business has not maintained good inventory records or has not updated its bills of materials with all relevant changes, it may encounter serious problems with the outputs of its MRP system. The problems could range from missing parts and excessive order quantities to schedule delays and missed delivery dates. At a minimum, an MRP system must have an accurate master production schedule, good lead-time estimates, and current inventory records in order to function effectively and produce useful information.

Another potential drawback associated with MRP is that the systems can be difficult, time consuming, and costly to implement. Many businesses encounter resistance from employees when they try to implement MRP. For example, employees who once got by with sloppy record keeping may resent the discipline MRP requires. Or departments that became accustomed to hoarding parts in case of inventory shortages might find it difficult to trust the system and let go of that habit.

The key to making MRP implementation work is to provide training and education for all affected employees. It is important early on to identify the key personnel whose power base will be affected by a new MRP system. These people must be among the first to be convinced of the merits of the new system so that they may buy into the plan. Key personnel must be convinced that they personally will be better served by the new system than by any alternate system. One way to improve employee acceptance of MRP systems is to adjust reward systems to reflect production and inventory management goals.

## Manufacturing Resource Planning

Manufacturing resource planning (MRP II) is a comprehensive type of planning for manufacturing companies. It is a sort of extension to the original material requirements planning (MRP) concept. It emerged in the 1980s to help companies deal with dynamic processes. Both of these, MRP and MRP II, are related to the enterprise resource planning (ERP) system, which is a top-level business information system that helps companies to plan better and work more efficiently.

Manufacturing resource planning may include various software tools as well as support processes. It is an overarching concept for business management. The tools may include master production schedules, advanced invoicing, production resources, inventory tools and more. The support processes may include contract management, shop floor data collection, sales analysis and more.

Through the use of diverse new technologies, companies can adjust how they work to improve productivity and efficiency. Inventory control systems are a good example — by aggregating big data and analyzing them for business intelligence, companies can reduce warehouse inventory levels, to save on maintenance cost. This is just one way that MRP works for businesses; another way involves improving supply chains as well as other parts of the production cycle.

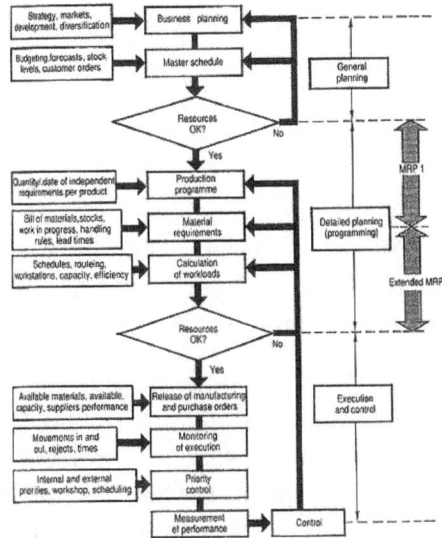

## Elements of Manufacturing Resource Planning

Manufacturing resource planning features important feedback from the production floor and relay both progress and delay. It integrates this information to all levels of schedule so that the next run is updated on a regular basis.

- Schedule of Resources– Inventory management and control includes a scheduling capability that concentrates on resources such as equipment, machinery, and raw materials, which are crucial in the production of the final goods. This feature is essentially where MRP II got its name from. MRP II allows for the generation of comprehensive and accurate data, which helps personnel gain tighter control of the manufacturing process.

- Batching Guidelines– Batching guidelines are integrated as it is an important element in the scheduling of resources. MRP software systems feature a host of batching rules. The most important elements however are,

- Lot for Lot– Lot for lot essentially means batches that perfectly match client orders. For example, a company will only make 20 finished products for Product A, to be followed by 10 of Product B. these batches are then automatically followed throughout the process to ensure that it matches the requirements of customers.

- Economic Batch Quantity (EBQ)– In the case of EBQ, the size of the batch is computed using a formula that significantly reduces cost through balancing between set up cost and cost of stock. See the image below for explanation.

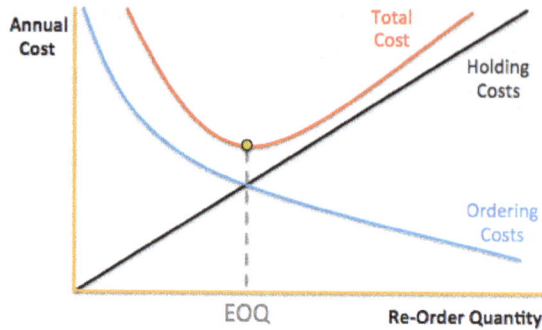

- Part Period Cover – Part Period Cover essentially means production of batches that matches the demand for a fixed period of time. Guidelines of making products on a weekly basis is an example of Part Period Cover.

- Software Extension Capabilities– Aside from inventory and resource control, there are different other programs that are included in MRP II. Some of these tools were designed to make the scheduling procedure more efficient. The MRP II may also include an option for Sales Ordering Processing. Manufacturing Resource Planning may also include stock recording and cost accounting programme, all of which are integrated into a company's main database system. There are also MRP that feature advanced planning capabilities that include:

  ◦ Computerized ordering of raw materials

  ◦ Hard and soft allocation

  ◦ Ideal versus present analysis of resource materials

  ◦ Minimum and maximum panning activities.

- Planning for Labour Capacity– MRP facilitates calculation of standard labor against the number of hours needed to meet daily, weekly, or monthly labor schedules. It automatically manages labor by means of category and qualifications.

- Accurate Data– Initial data typed in should be accurate in order for the MRP II to positively impact the manufacturing process. Errors in encoding information in the system results in a variety of problems to the business. Companies that develop and distribute MRP recommend users to carefully input data to achieve up to 98% or higher in terms of accuracy.

## Key Benefits of Manufacturing Resource Planning

- The MRP provides rich, detailed information that can be utilized by the company in core functions such as planning, management decision-making, and production.

- Reduce overall workload and maximum efficiency if the MRP system is managed correctly at all times.

- Data from MRP enables the management to plan ahead, and forecast how such planning can affect the overall profitability of the business.

- Planning with the help of MRP makes a company more efficient, thus avoiding unnecessary expenditures while maximizing profits altogether.

- MRP is an innovative and highly competitive tool that undeniably beats the traditional system of stock control and management.

- A fully-automated MRP software system easily and efficient formulates a production schedule which saves the company a large sum of money, time, and labor.

## Companies which should Implement MRP

The most ideal manufacturing business models that will greatly benefit from Manufacturing Resource Planning are those that feature a lot of assembly lines and base operates on the type of product to be sold. Even small businesses that offer both simple and complex products need to integrate MRP into their current operations to further improve production and increasing revenues as an end-result.

A well-managed and controlled Manufacturing Resource Planning software program can lead to the following positive results:

- Inventory levels are reduced, thus reducing inventory investment costs

- Successfully reduce amount of Work-in-Progress inventory

- Significantly improve customer experience and customer service by avoiding late orders

- Improvement in productivity during the manufacturing process

- Assists the company to adjusting to demand changes at a quicker and more efficient rate.

# Master Data Management

Master data management (MDM) is a comprehensive method of enabling an enterprise to link all of its critical data to a common point of reference. When properly done, MDM improves data quality, while streamlining data sharing across personnel and departments. In addition, MDM can facilitate computing in multiple system architectures, platforms and applications.

The benefits of master data management increase as the number and diversity of organizational departments, worker roles and computing applications expand. For this reason, MDM is more likely to be of value to large or complex enterprises than to small, medium-sized or simple businesses.

When companies merge, the implementation of MDM offers challenges, as various units consider the meaning of terms and entities intrinsic to their businesses. But, in mergers, master data management offers benefits, as it can help minimize confusion and optimize the efficiency of the new, larger organization.

For MDM to function at its best, personnel and departments must be taught how data is to be described, formatted, stored and accessed. Frequent, coordinated updates to the master data record are also essential.

## Issues

In the enterprise there are several systems managing the same data; the role of MDM is to centralize the data management to only one master copy of the data item which is then synchronized to all applications using the data. Using this approach, when referring to (for example) a customer within the enterprise, all systems are referring to the same customer.

There are basically two reasons why there are duplicated data which are inconsistent:

- The production systems within an enterprise, when implemented, have not been designed to be a part of larger set of production systems with which they should cooperate. Therefore, each system manages data on its own.

- The branches or departments of the company exist on their own without close cooperation with other departments. For example, the mortgage department deals with customers and manages the mortgage contracts. While the marketing department plans a promotion on mortgages. If the two departments do not cooperate (share the data), the marketing department may offer a mortgage to a customer who already has a mortgage. This is both a waste of money on the promotion as well as annoying to the customer.

- Company acquisitions or mergers are another example when an enterprise gets several parallel systems managing similar and sometimes overlapping data.

## Solutions

To handle the issues mentioned above, the common baseline for Master Data Management solutions comprises the following processes:

- Source identification- the 'system of record' needs to be identified first. If the same record is stored in multiple systems, the system which holds the most relevant copy (most valid, actual, or complete) of that record is referred to as a 'system of record'.

- Data collection- the data needs to be collected from various sources as some sources may attach a new piece of information, while dropping pieces which they are not interested in.

- Transformation - the transformation step takes place both during the input, while data are converted into a format for MDM processing, as well as on the output when distributing the master records back to the particular systems and applications.

- Data consolidation- the records from various systems which represent the same physical entity are consolidated into one record - a master record. The record is assigned a version number to enable a mechanism to check which version of record is being used in particular systems.

- Data duplication- often there are separate records in the company's systems, which in fact identify the same customer. For example, the bank may have a record identifying a customer while the bank's insurance subsidiary or department maintains a separate database of customers having a different record for the same customer. It is vital that these two records are duplicated and maintained as one master record.

- Error detection - based on the rules and metrics, the incomplete records or records containing inconsistent data should be identified and sent to their respective owners before publishing them to all the other applications. Providing erroneous data may compromise credibility of the company's MDM.

- Data correction - related to error detection, this step notifies the owner of the data record that there is a need to review the record manually.

- Data distribution/synchronization - the master records are distributed to the systems in the enterprise. The goal is that all the systems are using the same version of the record as soon as possible after the publication of the new record.

## Process

In the previous paragraphs, we have mentioned that each data record has to be assigned its owner or steward - a person who understands the data and is responsible for maintaining the record. The steward needs to be from the business side of the company. The reason, of course, is that only a business person understands the data and can make decisions about the data consolidation, updates, corrections and validity. On the other hand, the actual processing can be made available to either the business user via GUI or the IT department.

## References

- Methodologies-erp-systems-pmbok-8166: pmi.org, Retrieved 22 July 2018

- 5-types-erp-software-delivery-models: exactlly.com, Retrieved 29 June 2018

- Material-requirements-planning-mrp: inc.com, Retrieved 19 April 2018

- Manufacturing-resource-planning-mrp-ii-20604: techopedia.com, Retrieved 24 April 2018

- Importance-proper-manufacturing-resource-planning: manufacturing-software-blog.mrpeasy.com, Retrieved 24 April 2018

- Master-data-management: dataintegration.info, Retrieved 25 June 2018

# Permissions

# Index

www.ingramcontent.com/pod-product-compliance
Lightning Source LLC
Chambersburg PA
CBHW061957190326
41458CB00009B/2894